Affective Computing and Sentiment Analysis

TEXT, SPEECH AND LANGUAGE TECHNOLOGY

VOLUME 45

Series Editor

Nancy Ide, *Vassar College, New York*

Editorial Board

For further volumes:
http://www.springer.com/series/6636

Affective Computing and Sentiment Analysis

Emotion, Metaphor and Terminology

Edited by

Khurshid Ahmad
Trinity College, Ireland

 Springer

Editor
Khurshid Ahmad
Trinity College
School of Computer Science and Statistics
Dublin 2
Ireland
kahmad@scss.tcd.ie

ISSN 1386-291X
ISBN 978-94-007-3788-4   ISBN 978-94-007-1757-2 (eBook)
DOI 10.1007/978-94-007-1757-2
Springer Dordrecht Heidelberg London New York

Springer is part of Springer Science+Business Media (www.springer.com)

Acknowledgments

I must begin by acknowledging the patience and goodwill of the contributors of this book. Many of the authors traveled to Marrakesh to participate in the *Workshop on Emotion, Metaphor, Ontology, and Terminology (EMOT)* in August 2008 from places as far as the USA, Hong Kong and the EU. This Workshop was part of the bi-annual, EU-sponsored *Language Resources and Evaluation Conference (LREC)*. Selected papers from the Workshop were revised by the contributors during 2009–2010. Equally patient and helpful were my editors Jolanda Voogd and Helen van der Stelt of Springer Verlag. I am grateful to all the authors and editors. Last but not least, my grateful thanks to my Project Manager, Sangeetha Sathiamurthy at Springer's, for all the help she has provided during the final stages of the production of this book.

Papers presented in this book were reviewed by the following colleagues: Gerhard Budin (Universität Wien, Vienna), Ann Devitt (Trinity College, Dublin), Sam Glucksberg (Princeton University), Gerd Heyer (Universität Leipzig), Mícheál Mac An Airchinnigh (Trinity College, Dublin) Maria Teresa Musacchio (University of Padova), Maria Teresa Pazienza (University of Roma Tor Vergata),Margaret Rogers (University of Surrey, UK), Carl Vogel (Trinity College, Dublin), Tony Veal (University College, Dublin), and Yorick Wilks (University of Sheffield, UK). The whole collection was reviewed by anonymous referees selected by my editors Jolanda and Helen. My grateful thanks to all the named and anonymous referees.

My colleagues, Andrea Zemánková, and Chaoxin Zheng at Trinity College, helped in formatting the initial draft of this book. Thanks are due to my doctoral students, Aaron Gerow and Daniel Isemann, who formatted the final draft of this collection and helped in proof reading.

The support of Long Room Hub Project, Trinity College, Dublin, Ireland, is gratefully acknowledged.

Introduction: Affect Computing and Sentiment Analysis

Khurshid Ahmad

Many of the things we think about, actions we take, the way we react to stimuli, generate a feeling or subjective experience, for example, an emotion, or a mood. The generic term used in the twentieth century psychology and philosophy literature to denote such an emotion or mood is an old, Middle English (fourteenth century) word *affect*. The outward display of affect is manifested typically by facial expressions and postures by humans. Language plays a key role in the articulation of affect both in spoken language and in written text: more interestingly perhaps is the use of metaphors in the articulation of emotion, mood and sentiment. The use of metaphors to express affect is not restricted to the imaginative genre of fiction and can be found in specialist languages of science in general [11], and in natural [12] and biological sciences [17] in particular. Finance and economics have their fair share of metaphors for expressing affect.

Much in the vein of *doctrine of affects* in musical aesthetics, that music can be composed to arouse a variety of specific and involuntary emotions in the listener, behavioural psychologists have argued that an outward display of specific emotions can be precipitated by *framing* political or financial propositions in a particular way [8]. The metaphorical meltdown of the global financial systems in 2008 was precipitated, in part, by strong, negative sentiment about financial systems in news reports and editorials penned by journalists, in op-ed columns written by commentators and experts, in blogs by citizens, and in speeches by bankers and bank regulators alike. Equally important in the meltdown was the contribution of the actively promoted, positive sentiment in reports and comments on the so-called riskless investments; these texts comprised hyperbole like *collateral obligations* and inelegant coinages like *liars loans*.

The 2008 financial crisis, and many before that, shows how figurative language can be used to engender a false sense of security prior to the crisis and then constructs in the same language can be used to pacify the victims of the crisis afterward. The use of figurative language to describe the changes in financial markets has a long pedigree: In the seventeenth century we had *financial mania*, the eighteenth

K. Ahmad (✉)
School of Computer Science and Statistics, Trinity College, Dublin 2, Ireland
e-mail: kahmad@scss.tcd.ie

had *financial bubbles*, the nineteenth had *financial panic* only to be replaced by the emotive term *financial market crash* and the psycho-babblish *economic depression* in the first half of twentieth century [4]. The dominance of physical metaphors in the second half of the twentieth century begat the wave-like *economic recession*. Climatic metaphors were used in the twenty-first century, for example *credit freeze*, only to be replaced by the stronger-closer-to-the-facts terms like *credit squeeze* and *credit crunch*. These terms are adapted by scholars in finance and other disciplines, so, for example, *financial panic* has led to the term *banking panic* in modern finance literature [3, 13]. The creativity of experts in finance and their siblings on and around the trading floors continues apace: there are *haircuts* to describe investor losses [15] and the intriguing *dark pools* denoting a new type of exchange designed to address the problems that arise from the transparent nature of a typical stock exchange [9]!

The rolling global news, usually available digitally, and the ease with which opinions can be expressed about such news, is not limited to economics or finance, to politics or science, but now encompasses every walk of life. This new 'democracy', which allows almost anybody with access to a digital device (computers, mobile phones) to express their opinions, means a proliferation of vast amounts of digital text and speech. This large volume of news and views, often comprising emotion and mood, cannot be analysed for its subjective contents by humans within a reasonable time if at all. The search for, and the importance of, feelings and subjective experience articulated in language, has led to the coinage of the term *sentiment analysis*: a systematic, computer-based analysis of written text or speech excerpts, for extracting the attitude of the author or speaker about a specific topic. Indeed, the origin of the sentiment analysis lies in the mid-twentieth century political science, where pioneers like Harold Lasswell [10] and Philip Stone [14], developed *content-analysis* systems for analysing speeches by politicians and the manifesto of political parties. Stone created the *General Inquirer* system, and his dictionary of affect words, with 82 different categories of affect, is still used today [1, 16].

The goal of the computer scientists involved in building such *sentiment analysis* systems, is to identify and tabulate the judgement of the author/speaker or to discern how much or how little the author speaker values an object, event, state, or abstract idea. The longer term aim of such an analysis is to identify and tabulate what, if any, involuntary emotional reaction a reader/listener may have to an author/speaker. Indeed, sentiment analysis *aka* opinion mining, is used extensively in brand management and product promotion [2]. In economics and finance, sentiment analysis enthusiasts suggest that in addition to *fundamental analysis* (the analysis of the assets of an enterprise) and *technical* analysis (the analysis of the changes in the values of share or commodity prices), it is essential to use *sentiment analysis* – the inclusion of rumours in the market place related to an impending announcement of a break through or meltdown, in order to explain phenomenon that cannot be explained by fundamental and technical analysis. These phenomena include the technology induced booms and busts, like the dot com boom/bust, and spikes in commodity prices.

Language, as it appears to the contributors in this volume, is the key to sentiment analysis. Affect is articulated in text and speech through an exciting use of metaphors. Sam Glucksberg [5] has used the metaphors – *my spouse's lawyer is a shark* and *my job is my jail* – to explain how language is used to represent something as representative or suggestive of something else. For Andrew Goatly, the ubiquity of metaphor in everyday language and in special languages of science, technology, and the arts, suggests that 'far from being an anomaly', metaphors 'become' basic [6]. Cognitive scientists, like Jerry Hobbs, motivated by 'the computer metaphor of the mind' [7], have attempted to address the notions of *quantitative semantics* by, for example, looking at spatial relationships. Yorick Wilks, a pioneer of word-sense disambiguation and information extraction [18], links the enterprise of sentiment analysis to be a synthesis of work carried out in artificial intelligence, natural language processing, computational linguistics and content analysis (this volume).

In this collection of papers the contributions fall into three interacting categories – creation of metaphors, affect transfer in conversation, and sentiment analysis systems in economics, finance and marketing. The epilogue of this collection envisions future information extraction systems that may extract affect in free text and speech:

I. Language use and the creation of metaphors. Here we have Sam Glucksberg's contribution on the creation of new categories using metaphors, Andrew Goatly's on how metaphors are used in conceptualisation and in the expression of opinions, and Jerry Hobbs and Andrew Gordon's exploration of the semantic basis of metaphors. Carl Vogel synthesises the notions of metaphor formation and belief revision for arguing that metaphors and generics are species of the same kind – his paper also provides a link to the analysis of emotive language use.

II. Affect transfer in conversation and uncertainty. Alan Walington, John Barnden and colleagues describe an innovative approach to affect transfer in conversational situations and Rieks op den Akker and colleagues deal with detecting 'uncertainty in spoken dialogues'.

III. Sentiment analysis in economics, finance and marketing.

Economics and Finance: Maria Teresa Musacchio looks at metaphors used in economics in English and Italian and asks whether metaphors are universal or culture specific. Moshe Koppel and colleagues describe how supervised learning methods can be used to learn the association between polarity of financial news and key financial indicators. Khurshid Ahamd introduces the econometric notions of *return* and *volatility* for measuring changes in sentiment as articulated in (financial) news: these changes in sentiment are then correlated with important financial indicators for analysis and prediction.

Marketing and Product Reviews: Gerd Heyer and associates look at sentiment analysis in marketing. Marc Boullé together with Damien Poirier and colleagues, compares and contrasts the relative merits of machine learning and linguistic analysis techniques in predicting the opinions of reviewers on movie a blog site.

The *epilogue* of this collection of papers on has been written by Yorick Wilks. He argues that the interaction of contents analysis, artificial intelligence, particularly natural language processing, cognitive sciences, and information extraction may help us in understanding how metaphor is communicated by humans and may help us in building machines that can process metaphors and possibly sentiments.

References

1. Ahmad, Khurshid. 2008. Edderkoppspinn eller nettverk: News media and the use of polar words in emotive contexts. *Synaps* 21:20–36.
2. Candillier, Laurent, Frank Meyer, and M. Boullé. 2007. Comparing state-of-the-art collaborative filtering systems. International conference on Machine Learning and Data Mining MLDM 2007, Leipzig/Germany. Lecture Notes in Computer Science, 2007, Volume 4571/2007:548–562. Berlin: Springer.
3. Chari, V.V., and Ravi Jagannathan. 1988. Banking panics, information, and rational expectations equilibrium banking panics, information, and rational expectations equilibrium. *The Journal of Finance* 43(3):749-761, (July, 1988).
4. Galbraith, John K. 1958/1998. *The affluent society* – The fortieth anniversary edition. New York: Houghton Mifflin Company.
5. Glucksberg, Sam. 2001. *Understanding figurative language: From metaphors to idioms.* Oxford: Oxford University Press.
6. Goatly, Andrew. 1997. *The language of metaphors.* London: Routledge.
7. Hobbs, Jerry R.1990. Literature and cognition. Lecture notes, number 21, Center for the Study of Language and Information, Stanford, California.
8. Kahneman, Daniel, and Tversky, Amos. 2000. Choices, values, and frames. In *Choices, Values, and Frames*, eds. D. Kahneman and A. Tversky, 1–16. Cambridge, NY: Cambridge University Press.
9. Kuzman, G., Nevmyvaka, Y., Kearns, M., and , J.W. Vaughan. 2010. Censored exploration and the dark pool problem. *Communications of the ACM.* 53(5):99–107, (May 2010).
10. Lasswell, Harold D., Nathan Leites and Associates, eds. 1949. Language of politics: Studies in quantitative semantics. New York: George W. Stewart, Publisher, Inc. (Republished in 1965 by the MIT Press, Cambridge, MA).
11. Miller, A.L. 1984. *Imagery in scientific thought.* Boston, Basel, Stuttgart: Birkhaeuser.
12. Pullman, B. 1988. *The atom in the history of human thought.* Oxford: Oxford University Press.
13. Richardson, G., and W. Troost. 2009. Monetary intervention mitigated banking panics during the great depression: Quasi-experimental evidence from a federal reserve district border, 1929–1933. *Journal of Political Economy* 117(6):1031–1073, (December 2009).
14. Stone, Philip J., Dexter C. Dunphy, Marshall S. Smith, Daniel M. Ogilvie, and Associates. 1966. *The general inquirer: A computer approach to content analysis.* Cambridge, MA: The MIT Press.
15. Sturzenegger, F., and J. Zettelmeyer. 2008. Haircuts: Estimating investor losses in sovereign debt restructurings, 1998–2005. *Journal of International Money and Finance* 27(5):780–805, (September 2008).
16. Tetlock, Paul C. 2007. Giving content to investor sentiment: The role of media in the stock market. *Journal of Finance* 62:1139–1168.
17. Verschuuren, G.M. 1986. *Investigating the life sciences: An introduction to the philosophy of science.* Oxford: Pergamon Press.
18. Wilks, Y. 2008. On whose shoulders? *Computational Linguistics* 34(4):1–16.

Contents

Contributors

Rodrigo Agerri School of Computer Science, University of Birmingham, Birmingham B152TT, UK, R.Agerri@cs.bham.ac.uk

Khurshid Ahmad School of Computer Science and Statistics, Trinity College, Dublin 2, Ireland, kahmad@scss.tcd.ie

Rieks op den Akker University of Twente, Enschede, The Netherlands, infrieks@ewi.utwente.nl

John Barnden School of Computer Science, University of Birmingham, Birmingham B152TT, UK, J.A.Barnden@cs.bham.ac.uk

Cécile Bothorel France Telecom RD, TECH/EASY, 22300 Lannion, France, cecile.bothorel@orange-ftgroup.com

Marc Boullé France Telecom RD, TECH/EASY, 22300 Lannion, France, marc.boulle@orange-ftgroup.com

Jeroen Dral University of Twente, Enschede, The Netherlands, j.s.dral@student.utwente.nl

Michel Généreux Laboratoire d'informatique de Paris-Nord, Université Paris 13, 93430 Villetaneuse, France; Departamento de Linguística Geral e Românica, Faculdade de Letras da Universidade, de Lisboa Alameda da Universidade, 1600-214 Lisboa, Portugal, genereux@clul.ul.pt

Sam Glucksberg Princeton University, Princeton, NJ 08540, USA, samg@Princeton.EDU

Andrew Goatly Lingnan University, Tuen Mun, Hong Kong, goatly@ln.edu.hk

Andrew S. Gordon Institute for Creative Technologies, University of Southern California, Marina del Rey, CA 90292, USA, gordon@ict.usc.edu

Gerhard Heyer University of Leipzig, Natural Language Processing, Johannisgasse 26, D-04103 Leipzig, heyer@informatik.uni-leipzig.de

Dirk Heylen University of Twente, Enschede, The Netherlands, heylen@ewi.utwente.nl

Jerry R. Hobbs Information Sciences Institute, University of Southern California, Marina del Rey, CA 90292, USA, hobbs@isi.edu

Moshe Koppel Department of Computer Science, Bar-Ilan University, 52900 Ramat-Gan Israel, koppel@cs.biu.ac.il

Mark Lee School of Computer Science, University of Birmingham, Birmingham B152TT, UK, M.G.Lee@cs.bham.ac.uk

Patrick Mairif University of Leipzig, Natural Language Processing, Johannisgasse 26, D-04103 Leipzig, pmairif@informatik.uni-leipzig.de

Maria Teresa Musacchio Dipartimento di Lingue e Letterature AngloGermaniche e Slave, Università di Padova, Via Beldomandi 1, 35137 Padova, Italy, mt.musacchio@unipd.it

Émilie Guimier De Neef France Telecom RD, TECH/EASY, 22300 Lannion, France, emilie.guimier@orange-ftgroup.com

Thierry Poibeau Laboratoire d'informatique de Paris-Nord, Université Paris 13, 93430 Villetaneuse, France; Now at Lattice-CNRS, Ecole Normale Supérieure, 1 rue Maurice Arnoux, 92120 Montrouge, thierry.poibeau@ens.fr

Damien Poirier France Telecom RD, TECH/EASY, 22300 Lannion, France; Le Laboratoire d'Informatique Fondamentale d'Orléans, Université d'Orléans, Rue Léonard de Vinci, F-45067 Orleans Cedex 2, France, damien.poirier@univ-orleans.fr

Tim Rumbell School of Computer Science, University of Birmingham, Birmingham B152TT, UK, T.H.Rumbell@cs.bham.ac.uk

Katja Schöntag University of Hamburg, Marketing and Innovation, Von-Melle-Park 5, D-20146 Hamburg, Germany, schoentag@econ.uni-hamburg.de

Thorsten Teichert University of Hamburg, Marketing and Innovation, D-20146 Hamburg, Germany, teichert@econ.uni-hamburg.de

Carl Vogel Computational Linguistics Group, Intelligent Systems Laboratory, Department of Computer Science, O'Reilly Institute, Trinity College (University of Dublin), Dublin 2, Ireland, vogel@cs.tcd.ie

Alan Wallington School of Computer Science, University of Birmingham, Birmingham B152TT, UK, Alan.Wallington@gmail.com

Yorick Wilks University of Sheffield and Oxford Internet Institute, Sheffield, UK, yorick@dcs.shef.ac.uk

Chapter 1
Understanding Metaphors: The Paradox of Unlike Things Compared

Sam Glucksberg

1.1 Introduction

Metaphors pose thorny problems in diverse disciplines: rhetoric, philosophy, linguistics and psychology, among others. And, as with most thorny problems, people have vigorously expressed diametrically opposing views. For Nietzsche, metaphor is not merely a figure of speech. Rather, it is central to human thought, if not to humanity itself. Foreshadowing contemporary cognitive linguistics (cf Lakoff [22] and his colleagues), Nietzsche declared that "Knowing is nothing but working with the favorite metaphors... The drive toward the formation of metaphor is the fundamental human drive, which one cannot for a single instance dispense with in thought, for one would thereby dispense with man himself." [29] In contrast, many view metaphor as pernicious and misleading, at best. Hobbes, for example, asserted that absurd conclusions follow from "...the use of metaphors, tropes and other rhetorical figures instead of *words proper*...such speeches are not to be admitted." [19] (emphasis added). Why are metaphors "not to be admitted"? Because metaphors afford interpretations that clash with the literal and so words must be taken literally lest they lose their meanings. The protagonist of Haddon's novel about an autistic youth named Christopher Boone echoes Hobbes' view: "I find people confusing [because they] often talk using metaphors, such as *He was the apple of her eye, We had a pig of a day, They had a skeleton in the cupboard, The dog was stone dead*. I think [metaphor] should be called a lie because a pig is not like a day and people do not have skeletons in their cupboards and...imagining an apple in someone's eye doesn't have anything to do with liking someone a lot and it makes you forget what the person is talking about". Similes, on the other hand, are perfectly acceptable to Christopher: "He had a very hairy nose. It looked like there were two very small mice hiding in his nostrils. This is not a metaphor, it is a simile, which means that it really did look like there were two very small mice hiding in his nostrils. And a simile is not a lie, unless it is a bad simile" [18, p. 17].

S. Glucksberg (✉)
Princeton University, Princeton, NJ 08540, USA
e-mail: samg@Princeton.EDU

K. Ahmad (ed.), *Affective Computing and Sentiment Analysis*, Text, Speech
and Language Technology 45, DOI 10.1007/978-94-007-1757-2_1,
© Springer Science+Business Media B.V. 2011

1.2 The Metaphor Paraphrase Problem and the Priority of the Literal

Christopher's view of metaphor versus simile is extreme, relying, as Hobbes did, on an unqualified endorsement of literal meanings irrespective of context (and, of course, of actual language use). Furthermore, it ignores an important property of metaphors and similes. In most cases, any given metaphor can be paraphrased as a simile, and vice-versa. For example, a common interpretation of the metaphorical assertion "My surgeon was a butcher" is that the surgeon in question is not skilled and precise, to say the least. The interpretation of this metaphor would not change substantially if the metaphor were to be replaced by its corresponding simile, "my surgeon was *like* a butcher." In contrast, literal categorical assertions, such as "A man is a human being", cannot be paraphrased as a comparison. As Carston put it, "....it makes no .... sense to say [of a man] that he is *like* a human being, given that he *is* one" [3, p. 358].

Aristotle [1] characterized metaphors as comparisons, and specifically as involving two *unlike* things compared. As we have seen, similes can be paraphrased as metaphors, e.g., *lawyers are like sharks* can be expressed as *lawyers are sharks*. Literal comparisons, in contrast, involve two like things compared, yet they cannot be paraphrased as categorical assertions, e.g., *coffee is like tea* cannot be paraphrased as *coffee is tea*. That two unlike things can be expressed as an identity relation while two like things cannot is a paradox – the paradox of two unlike things compared. This special property of metaphors and their corresponding similes is an important problem to be resolved. What is it about metaphorical categorical assertions that afford paraphrasing as similes (and vice versa) while literal categorical assertions do not? Following Aristotle [1], contemporary writers on metaphor such as Gentner and Wolff [5] and Searle [28] finessed the paraphrase problem by claiming that metaphors are implicit comparisons. Hence, not only *can* metaphors be paraphrased as comparisons, they *must* be treated as such in order to be understood. This claim was not made in the interest of solving the paraphrase problem. Indeed, the paraphrase problem was not even explicitly recognized as a problem. Instead, the ability to paraphrase metaphors as simile was taken to justify the standard pragmatic view of metaphor comprehension [28].

On this view, literal interpretations have unconditional priority. Whatever the context and whatever a speaker's intention might be, people always derive a literal interpretation of all utterances. If the derived literal interpretation fails to make sense in context, then *and only then* are alternative non-literal interpretations attempted. This step by step process is invoked to account for all non-literal interpretation, including such disparate forms as metaphors, metonymy, idioms, irony and indirect requests, among others. Thus, upon hearing a speaker say "can you pass the salt?", the initial interpretation must be that this is a question about one's ability to pass the salt, to which a literal response would be to simply say "yes I can". But this is, of course, ridiculous. According to Grice's [17] cooperative principle, people in conversations expect one another to be sincere, truthful, informative and relevant.

The speaker knows that the addressee is *capable* of passing the salt, so the question cannot be about whether or not the addressee is able to pass the salt. Therefore, this literal interpretation is rejected and the addressee now infers that the speaker intended the utterance as a request, not as a question about ability.

Intuitively, this seems rather far-fetched, and there is abundant evidence that this is not just far-fetched, it is patently wrong. People do not always initially interpret every utterance in its literal sense [7]. People also do not, as required by the step-by-step view, take longer to arrive at non-literal interpretations than literal ones, [2]. More importantly, people do not need to reject a literal meaning before deriving non-literal meanings [13]. Language processing, in general, is automatic in the sense that we cannot turn off our language processor – we cannot refuse to understand. As Miller and Johnson-Laird put it, understanding "occurs automatically without conscious control by the listener. . .loss of control over one's language comprehension device may correspond to knowing a language fluently" [24, p. 166]. Linguistic input automatically triggers semantic and syntactic analyses that generate literal sentence meanings [4]. But linguistic input also automatically triggers pragmatic and conversational analyses that generate non-literal meanings as well [8–11]. In short, whether utterances are intended as indirect requests, idioms, metaphors, ironies or sarcasm, literal meanings do not have priority, and are not arrived at more quickly than are non-literal meanings.

1.3 Understanding Metaphors: Comparison or Categorization?

Given that literal and non-literal meanings are generated automatically and in parallel [23], the question of *how* metaphorical meanings are generated remains. Are metaphors understood via a comparison process, or are they understood directly as categorical assertions? Consider, first, how comparisons might be understood. Assertions such as *X is like Y* can be understood via two quite different sets of operations: by feature matching or by categorization. Feature matching involves identifying features of X and of Y, and then determining which features of X and Y match. Given the statement *lemons are like oranges*, a subset of features of lemons might be *yellow, oblong in shape, juicy, acidic mouth-puckering taste, have seeds*. A subset of features of oranges might be *orange, round, juicy, acidic mouth-puckering taste, have seeds*. The features *juicy, acidic taste and have seeds* match, and so when asked, how are lemons and oranges alike, one would answer that both are juicy, taste acidic and have seeds. This minimally captures the similarity between lemons and oranges, but the answer would suffice.

An alternative to feature matching is categorization, which involves finding the nearest available category that subsumes both *X and Y*. In the case of lemons and oranges, one answer to the question of how lemons and oranges are alike would be that they are both citrus fruits. Via this single operation, all the relevant and properties and none of the irrelevant properties that are shared by lemons and oranges can be generated. Not only is this simpler than feature matching, it is also more

Fig. 1.1 Cross categorization of lawyer and shark

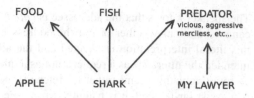

informative by, among other things, invoking other members of the citrus fruit category, such as grapefruits, to the mental representations of lemons, oranges and the properties that they share. It is thus not surprising to observe that the categorization answer is by far the most likely one when people are asked how lemons and oranges are alike. It is also the answer provided for this item when it appears in the similarities sub scale of the Wechsler Adult Intelligence Scale [31].

Categorization works quite well as a strategy to generate the basis for literal similarities. Consider the following, listed in order of increasing difficulty (after Wechsler, [31]). How are these pairs alike?

> *Oranges* and *Lemons*? Both fruits.
>
> *Oranges* and *Steak*? Both foods.
>
> *Oranges* and *Insects*? Both alive?
>
> *Lawyers* and *Sharks*? Both. . .Sharks?

This last comparison may not be a literal one, and indeed it is reflected in a fairly conventional metaphor, *My lawyer is a shark*. But how can a lawyer be a member of the category, *sharks*? As Fig. 1.1 illustrates, concepts can be categorized in any number of ways: *apples* and *sharks* are both foods, *sharks* and *lawyers* can both be predators in that both can be vicious, aggressive, merciless, etc. How might the category of such predators be named?

1.4 How Novel Categories Can Be Named: Dual Reference

The answer to this naming question also provides the solution to the metaphor-simile paraphrase puzzle. The word *shark* can refer literally to the marine creature that swims in the sea, has teeth, rough skin without scales, etc., This literal creature is the referent when we use a simile such as *my lawyer is **like** a shark* (See Fig. 1.2). But the word *shark** can also refer to the superordinate category of predatory creatures that subsumes both lawyers and literal sharks as category members.[1] This abstract category is the referent when we use a metaphor, such as *my lawyer is a shark**. The word form *shark* is thus not one word but two – one referring to the literal shark, the other, *shark**, referring to a category of predatory creatures.

---

[1] Following Carston [3], the superordinate category name will be denoted via a star, e.g., *shark**.

Fig. 1.2 Categorization
of lawyer and shark

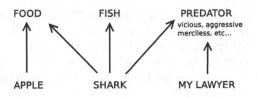

"Shark" used as a metaphor vehicle refers to a type
of thing, whereas used literally, it refers to an
actual shark.

"Shark", then, affords reference at two levels of abstraction. It thus affords dual reference.

Is dual reference unique to metaphors, or is it a general referring strategy that is used whenever a superordinate category has no name of its own? Consider classifier languages such as Hopi and American Sign Language. Classifier languages in general lack superordinate category names. In such languages, people use the name of a prototypical exemplar of a category as a name for that category. In Hopi, a native American South Western language, the cottonwood tree is the prototypical and most common deciduous tree. Because there is no word for the category "deciduous trees", Hopi speakers refer to any and all such trees as "cottonwoods". However, if there is a need to distinguish between cottonwoods and other trees, then people use "real-cottonwood" to refer to, of course, the real cottonwood [30]. American Sign Language (ASL) is a signed classifier language, and ASL users also use dual reference to refer to categories that have no names of their own. One such category is *furniture*, which can be referred to by using basic-level object signs that are prototypical of the furniture category. Newport and Bellugi [25, p. 62] cite the example of an ASL user who signed *house-fire [+] lose all **chair-table-bed, etc., but one left, bed***. In English, this might have been expressed by saying *I lost my furniture in the house fire, but there was one thing left, the bed*. In ASL, the three-word expression *chair-table-bed etc.* is signed quickly and serves as the category name, as clearly indicated by the end of the signed assertion, *one left, bed*. As in Hopi and as in metaphor, a word, in this case *bed*, has dual reference: as part of the three-word list, it refers to the category of furniture, and in that same sentence, to the individual category exemplar, *bed*.

But we don't need to look to classifier languages to find examples of dual reference. Dual reference is used in languages, including English, that have an abundant supply of names for common, familiar categories. In all language communities there are occasions when people need to refer to categories that do not yet have names of their own. When such an occasion arises, one strategy is to use the name of a prototypical exemplar of the as-yet unnamed category as the category name. Brand names for new products provide a classic example. One of the first brands of paper tissues was "Kleenex", and so, lacking a single-word name for the category "paper tissues", the Kleenex brand name became Kleenex*, the new category name. This dual reference now made it possible to paraphrase categorical assertions as comparisons: Referring to a paper tissue manufactured by Scott paper company, one can say "Scotties are Kleenex*" or "Scotties are like Kleenex". Similarly, the name for the category "Vacuum cleaner" in the UK is "Hoover", and so one could say that Mieles are or are like Hoovers. The new category name, Hoover*, even became the verb for the activity of vacuuming, as when a little girl, trying to get her turn to vacuum when her brother was playing with the machine said "Boys don't Hoover, only girls do!"

Dual reference is also seen in apparent tautologies, as in "boys will be boys*". The first boy refers to male individuals; the second to the category of people who behave (usually badly) as boys are reputed to behave. Here, too, dual reference makes category to comparison paraphrases permissible: one could say either "boys will be boys*" or "boys will be like boys", although the former seems more apt. One other kind of expression demonstrates the ubiquity of dual reference: the use of emblematic entities, entities that epitomize a category, as a name for the category itself. In the 1988 US presidential election, the democratic candidate for vice-president, Lloyd Bentsen, belittled his opponent by saying, "you, sir, are no John Kennedy". This was of course, literally true. Dan Quayle was not John Kennedy. But this was not the intended meaning. The intended meaning was that Dan Quayle did not belong to the category of admired political figures epitomized by President Kennedy. A contemporary example might be a supporter of Barack Obama in the 2008 US election saying "Obama is our John F. Kennedy" or, perhaps, an opponent saying "Obama is no JFK". And here too, categorical assertions can be paraphrased as comparisons and vice-versa.

1.5 Understanding Metaphors and Similes

As in these prosaic examples, the different forms of a metaphor – the categorical ISA and the simile "like" – have distinct referents. The categorical ISA form refers to the abstract metaphorical category, e.g., the metaphorical *shark*. The "like" form refers to the concrete exemplar of that category, the fish we call a shark. Figure 1.3 illustrates this distinction. Now, if the literal and metaphorical shark have distinct referents, then the mental representations of these two referents should be different. As illustrated in Fig. 1.4, The literal *shark* has all of the properties of the metaphorical *shark**, but it also has metaphor-irrelevant properties such as "can swim", "has fins", "leathery skin", "gills", etc. These "literal" properties are more likely to come to mind when the simile is used than when the metaphor is used, and when this happens, these metaphor-irrelevant properties must be filtered out [6, 26]. This implies that similes should be more difficult to understand than metaphors. More importantly, people's interpretations of metaphors and their corresponding similes should differ systematically – sometimes subtly so, but sometimes quiet drastically so.

With respect to the first implication, Johnson [20] examined the difference in time to understand metaphors in context, and found that similes did indeed take more time to understand and integrate into a text than did metaphors. With respect to differences between metaphors and similes, there is general agreement that metaphors

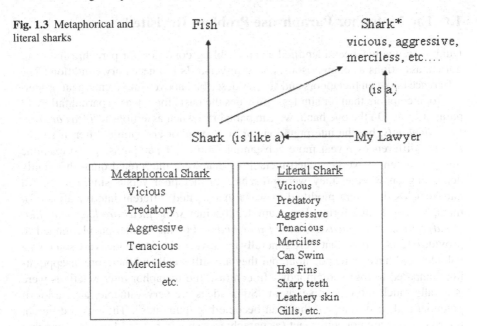

Fig. 1.3 Metaphorical and literal sharks

Fig. 1.4 Hypothetical representations of metaphorical and literal sharks

are somehow richer and more striking than similes [27], but there is no direct evidence for this – it relies primarily on intuition. One factor that may be responsible for this intuition is that metaphors are more "metaphorical" than similes in that they generate more emergent properties than do similes. An emergent property is one that is generated by two or more concepts, neither one of which has the property in question. For example, for the noun–noun combination "blind lawyer", a common emergent property is "courageous" – a property that is not ordinarily associated with either the concept "blind" or "lawyer." [21].

To see if metaphors give rise to more emergent properties than do similes, we asked college students to paraphrase metaphors in either metaphor or simile form. Literal properties are those that are normally associated with the metaphor vehicle and applicable to the metaphor topic. Emergent metaphorical properties are those that are not normally associated with the metaphor vehicle and are also applicable to the metaphor topic. For the metaphor *some ideas are diamonds**, a literal, basic-level property would be *very valuable*. In contrast, *creative* is not a property of the literal diamond (the gemstone) but can apply to the metaphorical diamond and, of course, to the topic *some ideas*. When paraphrasing metaphors, many more emergent properties were listed than were literal properties, e.g., *some ideas are insightful, some ideas are creatively very unique*. In contrast, when paraphrasing similes, the reverse was true – many more literal rather than emergent metaphor properties were listed, e.g., *rare and desirable, so interesting that they shine and glitter, very valuable*. In this sense, metaphors are seen as more metaphorical than similes [14].

1.6 The Metaphor Paraphrase Problem Revisited

Dual reference has been identified as an enabling condition for paraphrasing categorical assertions as comparisons, and vice-versa. Is it a necessary condition? The difference between metaphors and similes described above – more emergent properties for metaphors than for similes – provides the basis for at least a partial failure of paraphrasing. On the one hand, we can paraphrase such assertions as *ideas are (are like) diamonds*, but the interpretations of the metaphor and simile differ, if subtly. Sharper differences reveal more substantial failures of paraphrasing. To examine this phenomenon, we generated metaphors that were interpreted quite differently depending on whether they were presented in metaphor versus simile form. All had vehicles that were modified in ways that invited different interpretations for metaphor and simile forms. For example, consider *some ideas are like small diamonds* versus *some ideas are small diamonds**. For the simile, people tended to provide interpretations that were generally negative, as in "Some ideas are somewhat valuable and have some potential, but they are still small and therefore disappointing compared to bigger diamonds". In contrast, the metaphor interpretations were generally much more positive, as in "Some ideas are very valuable, have a lot of potential and, if developed, they can become big diamonds". These two different interpretations are not equivalent (as paraphrases should be) and I would argue that this is because the referring expression "small diamonds" does not afford dual reference in this context. The literal "small diamonds", which are not nearly as good as big diamonds, is not a member of the metaphorical category "small but still very valuable diamonds", as indicated by the different interpretations given to the simile versus the metaphor.

In this example, the evaluative nature of the metaphor and simile differs, but the content remains substantially the same. More serious violations of paraphrasability involve changes in the contents of metaphors and similes. Consider the differences in interpretations of the metaphor and simile *my lawyer was an old shark** versus *my lawyer was like an old shark*. For the metaphor, the predicate *old shark* was generally understood as referring to an old pro, in this case one who is competent, aggressive, experienced. This is a positive evaluation with correspondingly positive properties. For the simile, the same noun phrase is understood quite differently. A literal old shark is an old fish, and (alas) brings to mind a stereotype of old people. In this case it is a negative one, with correspondingly negative properties such as ineffectual, weak and toothless. This metaphor and its corresponding simile differs both in evaluation and content, but not in the category membership of the predicate referents: both in the metaphor and simile, the referent is an animate being or person – one who is competent, the other enfeebled.

In other cases, even the referent kind can differ between metaphor and simile, as with *his job was a secure jail** versus *his job was like a secure jail*. For the metaphor, participants generated paraphrases such as "His job was confining, but his income was guaranteed and he knew that he could not get fired." This is, on balance, positive, with appropriately negative and positive properties. For the simile, we found paraphrases such as "His job was confining, like a trap or a dead end that he found impossible to get out of." There is some overlap in the contents of the metaphor and simile paraphrases, but the simile is unambiguously negative with contents such as "trap" and "dead end" to complement "confining".

These findings suggest that dual reference is not just an enabling but a necessary condition for metaphors and similes to be paraphrases of one another. Perhaps a more important implication is that dual reference may not be a necessary condition for metaphors – indeed, that we can have metaphors that not only differ from their corresponding similes, but that cannot be paraphrased as similes at all. What sort of metaphor resists paraphrasing entirely? One kind of metaphor that behaves in this way is one in which the vehicle has no literal referent available, thus precluding dual reference, and hence also precluding paraphrasing as a simile. Consider *my lawyer was a well-paid shark** versus *my lawyer was like a well-paid shark*. The metaphor has a sensible interpretation. The lawyer that we first met as an aggressive predator is still an aggressive predator, and is well paid to be so. But what is the referent for the simile predicate? Real sharks are not entities that can be paid, let alone well-paid, and so the simile paraphrase of the metaphor just doesn't work. Admittedly, the shark example and others like it were created for our experiments. Are there examples from naturally occurring expressions that behave this way?

The media provide countless examples of metaphors that make no sense as similes because the referents of the metaphor vehicles simply do not literally exist. The worlds of commerce and finance provide abundant examples. When Enron failed spectacularly, it became emblematic of scandalous and disastrous economic failures. Predictably, the concept "Enron" became available as a metaphor for any such failures, including future ones. And sure enough, only a day or two after Enron's collapse, the New York Times ran the headline "Will Worldcom be the next Enron*?" This metaphor cannot felicitously be paraphrased as a simile; Worldcom cannot be like "the next Enron" because that entity does not yet exist. Worldcom can only *be* the metaphorical Enron. Similarly, the banking firm Lehman brothers could only *be* the next Bear Stearns (a firm that had collapsed earlier) until, unfortunately for Wall Street, Lehman Brothers actually became the next Bear Stearns. Once that happened then, and only then, could we say that Lehman Brothers was *like* Bear Stearns. And as usual, politics and government provide their own share of such examples, as when serious questions were raised about a state election in Florida. Referring to the voting controversies and miscounts in the 2004 presidential election in Florida, a state official monitoring voting in 2006 warned that ". . .unless we do something now, Florida is headed toward being the next Florida*" [16]. Here, as in the examples above, the literal "Florida" can only *be* a metaphorical "Florida" because the "next Florida" does not yet exist. If that came to pass, then one could say "Florida is like Florida", just as we can say "boys will be like boys" [15].

1.7 Comparison Versus Categorization Revisited

Dual reference not only makes it possible to paraphrase metaphors as similes, it seems to be necessary. However, dual reference does not seem to be necessary for metaphors – at least for those metaphors whose vehicles have no literal referent available. In such cases, metaphors cannot be adequately paraphrased as similes, if they can be paraphrased at all. This raises the issue of how metaphors are understood: via comparison, or via categorization? One possibility is that when a metaphor vehicle affords dual reference – to an abstract metaphorical category and to a literal exemplar of that category – then the process would depend on the metaphor form. If in metaphor form, then categorization would be most likely and appropriate. If in simile form, then comparison would be indicated. However, comparison could be accomplished either by feature matching, or by categorization (as in the oranges and lemons – both citrus fruits example). If, however, dual reference is not available, (as in the well-paid shark, the next Enron and other such examples) then the metaphor would have no simile counterpart. In such cases, categorization would be the only alternative.

1.8 Conclusions

Traditional theories of metaphor comprehension assume that literal meanings have unconditional priority. Literal meanings are always generated, and it is only when a literal meaning fails to make sense in context do people seek an alternative, non-literal interpretation. Given a false literal statement such as "my lawyer was a shark", people first generate and then reject the literal interpretation (that the lawyer in question is a fish) and then seek an alternative figurative interpretation by converting the false metaphor into a true simile, *my lawyer was like shark*. This move is justified on the grounds that metaphors can usually be paraphrased as similes and vice-versa, without appreciable changes in meaning. We review the evidence and conclude that in every relevant respect, the literal does not have priority. Metaphors are understood as quickly as comparable literal expressions, and understanding metaphors is not dependent on a literal meaning to be unacceptable in context. Like any linguistic expression, metaphors are processed automatically and in parallel with any literal meanings that may be available [8–11, 23].

Can metaphors be understood directly as category assertions, or must they be treated as comparison statements? One answer to this question comes from the observation that literal comparison statements, such as *oranges are like lemons*, can and most often are understood via categorization: oranges and lemons are alike in that both are citrus fruits [10, 12]. The same holds true for similes such as *lawyers are like sharks*. Lawyers and sharks are alike in that both belong to the category of aggressive, predatory creatures that sharks exemplify. Because that category has no name of its own, the name of the emblematic category exemplar is used as the name of the category itself, Shark*, via the referring strategy of dual reference. A single word form can refer either to the literal exemplar of the category (the fish we call a shark) or the more abstract category of predatory creatures, Shark*, which would include both the fish and the lawyer. Dual reference is what makes paraphrases of metaphors and similes possible. Furthermore, we found that dual reference is not only an enabling condition for such paraphrases, but a necessary one. This resolves the paradox of unlike things compared and also solves the metaphor paraphrase problem – not by assuming that metaphors are, deep down, similes, but by identifying the mechanism that makes paraphrases possible.

Dual reference is not, however, a necessary property of metaphors, only of those metaphors that can be paraphrased as similes. When does paraphrasing fail? We identified two kinds of dual reference failures – and hence paraphrasing failures. In the first, the predicates of the metaphor and simile differ sufficiently to render paraphrases non-equivalent, as in *his job was a secure jail** versus *his job was like a secure jail*. In the metaphor, *secure jail** refers to a job that is confining but guaranteed. In the simile, the literal secure jail refers to something like a maximum security prison, a dead end that one cannot get out of. The second type of dual reference failure occurs when the predicate of a simile does not exist, and hence it cannot be referred to by the simile predicate, as in *my lawyer was a well-paid shark**. Fish cannot be paid, and so a well-paid shark can only be a metaphorical one. The assertion *my lawyer was like a well paid shark* just doesn't make sense.

We conclude that metaphors can be understood directly via categorization, and usually can also be understood via comparison, but not in all cases. The major theoretical implication of this conclusion is clear. Because some expressions can work only as metaphors, and some metaphors change in meaning when paraphrased as similes, theories of metaphor that assume the equivalence of metaphors and similes are incomplete, if not fatally flawed. So we can now say to our autistic savant Christopher Boone, metaphors are not lies, and they are not similes, either.

References

1. Aristotle. 1946. *Rhetoric* (trans: Roberts, R.). Oxford: Oxford University Press.
2. Blasko, D. G. and C. M. Connine. 1993. Effects of familiarity and aptness on metaphor processing. *Journal of Experimental Psychology: Learning, Memory and Cognition* 19:295–308.
3. Carston, R. 2002. *Thoughts and utterances: The pragmatics of explicit communication.* Oxford: Blackwell.
4. Fodor, J. A. 1983. *The modularity of mind.* Cambridge, MA: Bradford Books.
5. Gentner, D. and P. Wolff. 1997. Alignment in the processing of metaphor. *Journal of Memory and Language* 37:331–355.
6. Gernsbacher, M. A., B. Keysar, R. R. Robertson, and N. K. Werner. 2001. The role of suppression in understanding metaphors. *Journal of Memory and Language* 45:433–450.
7. Gibbs, R. 1986. Skating on thin ice: Literal meanings and understanding idioms in conversation. *Discourse Processes* 9:17–30.
8. Gibbs, R. 1994. *The poetics of mind: Figurative thought, language, and understanding.* New York: Cambridge University Press.
9. Giora, R. 2003. *On our mind: Salience, context, and figurative language.* New York: Oxford University Press.
10. Glucksberg, S. 2001. *Understanding figurative language: From metaphors to idioms.* New York: Oxford University Press.
11. Glucksberg, S. 2004. On the automaticity of pragmatic processes: A modular proposal. In *Experimental pragmatics*, eds. I. A. Noveck and D. Sperber, 72–93. New York: Palgrave Macmillan.
12. Glucksberg, S. 2008. How metaphors create categories – quickly. In *The Cambridge handbook of metaphor and thought*, ed. R. Gibbs, 67–83. Cambridge: Cambridge University Press.

13. Glucksberg, S., P. Gildea, and H. A. Bookin. 1982. On understanding nonliteral speech: Can people ignore metaphors? *Journal of Verbal Learning and Verbal Behavior* 21:85–98.
14. Glucksberg, S. and C. Haught. 2000a. On the relation between metaphor and simile: When comparison fails. *Mind & Language* 21:360–378.
15. Glucksberg, S. and C. Haught. 2000b. Can Florida become *like* the next Florida? When metaphoric comparisons fail. *Psychological Science* 17:935–938.
16. Goodnough, A. 2004. Lost record of vote in '02 Florida race raises '04 concerns. *New York Times*, April 28, p. B4.
17. Grice, H.P. 1975. Logic and conversation. In *Syntax and semantics: Vol. 3. speech acts*, eds. P. Cole and J. Morgan, 41–58. New York: Academic Press.
18. Haddon, M. 2003. *The curious incident of the dog in the night-time*. New York: Doubleday.
19. Hobbes, T. 1929. *Hobbes's Leviathan*. Reprinted from the edition of 1651. With an essay by the late W. G. Pogson Smith. Oxford: Clarendon Press.
20. Johnson, A. T. 1996. Comprehension of metaphors and Similes: A reaction time study. *Metaphor and Symbol* 11:145–159.
21. Kunda, S., D. T. Miller, and Claire, T. 1990. Combining social concepts: The role of causal reasoning. *Cognitive Science* 14:551–577.
22. Lakoff, G. 1980. The metaphorical structure of the human conceptual system. *Cognitive Science* 4:195–208.
23. McElree, B. and J. Nordlie. 1999. Literal and figurative interpretations are computed in equal time. *Psychonomic Bulletin & Review* 6:486–494.
24. Miller, G. A. and P. Johnson-Laird. 1976. *Language and perception*. Cambridge, MA: Harvard University Press.
25. Newport, E. L. and U. Bellugi. 1978. Linguistic expressions of category levels in a visual-gesture language: A flower is a flower is a flower. In *Cognition and categorization*, eds. E. Rosch and B. B. Lloyd, 49–71. Hillsdale, NJ: Erlbaum.
26. Newsome M. R. and S. Glucksberg. 2002. Older adults filter irrelevant information during metaphor comprehension. *Experimental Aging Research* 28(3):253–267.
27. Ortony, A. 1979. Beyond literal similarity. *Psychological Review* 86:161–180.
28. Searle, J. 1979. Metaphor. In *Metaphor and Thought*, ed. A. Ortony, 92–123. New York: Cambridge University Press.
29. Schrift, A. D. 1990. *Nietzsche and the question of interpretation: Between hermeneutics and deconstruction*, 127–128. New York: Routledge.
30. Trager, G. L. 1936–1939. 'Cottonwood-Tree': A south-western linguistic trait. *International Journal of American Linguistics* 9:117–118.
31. Wechsler, D. 1958. *The measurement and appraisal of adult intelligence*, 4th ed. Baltimore, MD: Williams & Williams Co.

Chapter 2
Metaphor as Resource for the Conceptualisation and Expression of Emotion

Andrew Goatly

2.1 Background

Cognitive linguistic (CL) accounts of metaphor were popularised by Lakoff and Johnson [6], and are also associated with other scholars such as Turner, Sweetser, Gibbs, Steen, Kövecses, Radden and Barcelona. They stress the ubiquity and inescapability of metaphor in thought and language, and also recognise that the metaphors we use form mental structures or schemata realised by lexical sets, known variously as conceptual metaphors, root analogies or metaphor themes. In this paper I will use the latter term. These metaphor themes involve mappings between sources (vehicles) and targets (topics, tenors) and are traditionally labelled in small caps by the formula, TARGET IS SOURCE.

In the linguistics tradition where Lakoff was nurtured, there has been a tendency to intuit metaphor themes without much lexical evidence for their importance, and so to reach doubtful conclusions about, for example, the conceptualisation of emotions [1, p. 95]. To remedy this ad hoc intuitive approach, I undertook research to establish in a more principled way the important metaphor themes for English.[1] The somewhat arbitrary double criteria I used are: (1) To count as significant metaphor themes should be realised by at least 6 lexical items, found in a dictionary of contemporary English; (2) There should be at least 200 tokens of this joint set of lexical items with the relevant metaphorical meaning in the Cobuild Bank of English database. The website 'Metalude' [7] (Metaphor at Lingnan University Department of English) is the result of these endeavours.[2] It includes an interactive database of 9000+ English metaphorical lexical items, grouped by metaphor theme, and provides the data for this paper.

---

[1] The research was funded by the Research Grants Council Hong Kong SAR, reference LC3001/99H.

[2] http://www.ln.edu.hk/lle/cwd03/lnproject_chi/home.html. User id: ⟨user⟩, password: ⟨edumet6⟩.

A. Goatly (✉)
Lingnan University, Tuen Mun, Hong Kong
e-mail: goatly@ln.edu.hk

Section 2.2 is an overview of the ways in which emotion in general and specific emotions are metaphorically conceptualised in English, with a brief aside on anger in particular. Section 2.3 explores how English metaphor themes and their lexis contribute to the expression of emotion/evaluation.

2.2 Metaphorical Conceptualisation of Emotions in English

The second section falls into two parts: the first delineates the cluster of metaphor themes for the conceptualisation of emotion; and the second the question of the ambiguity between conceptualisation and expression of emotion, especially in cases where one describes one's own emotions in the first person.

2.2.1 Conceptualisation of Emotion

The major metaphor themes for conceptualising emotion in English can be organised in four loose hierarchies. The most important grouping (Fig. 2.1) has at the top of the hierarchy EMOTION IS SENSE IMPRESSION, with the remaining members of this group directly or indirectly dependent upon it. The important theme EMOTION IS WEATHER relates to all the sense impressions, except smell.

The second, much simpler, group depends upon the joint themes EMOTION IS FLUID and EMOTION IS MOVEMENT, or the movement of fluids (Fig. 2.2). It might be possible to see an experiential connection between these two groups: WEATHER involves MOVEMENT of FLUIDS (air and water); and EMOTION IS EXPLOSION can be linked to EXPRESSION IS OUTFLOW (of gas).

The third group uses space to indicate relationship. It too may be connected to the first group since PROXIMITY, especially in early childhood, is associated with

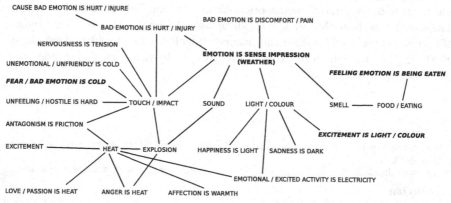

Fig. 2.1 Sub-themes for EMOTION IS SENSE IMPRESSION

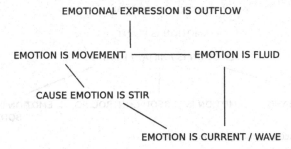

Fig. 2.2 EMOTION IS MOVEMENT-EMOTION IS LIQUID and their associated themes

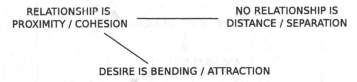

Fig. 2.3 RELATIONSHIP IS PROXIMITY/COHESION and its related metaphor

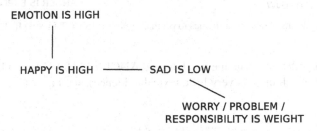

Fig. 2.4 EMOTION IS HIGH and its related metaphor themes

WARMTH, relating it to AFFECTION IS WARMTH in particular and EMOTION IS TOUCH more generally (Fig. 2.3).

The fourth group concerns orientational metaphor based on the vertical axis, by which emotions in general and happiness especially are conceived as being high (Fig. 2.4).

There remain, in Metalude, a number of miscellaneous metaphor themes which do not seem to form a systematic group, beyond the fact that EMOTION is concretised as a MINERAL, and then animised as various kinds of living thing – PLANT, ANIMAL (HUMAN), a human who is a PERSON CONTROLLED, and parts of a human, BODY PART/BODY LIQUID (Fig. 2.5). The latter is probably the vestige of medieval medicine, the doctrine of the four humours. The remaining source, DISEASE, might link with EMOTION IS SENSE IMPRESSION in group 1.

This kind of lexicological work shows that the subgroups are often cross-linked to form larger webs and schematic interactions, a complexity that Grady has attempted to remedy with his notion of primary metaphor, though at the cost of richness of imagery and psychological force. To indicate how metaphor themes

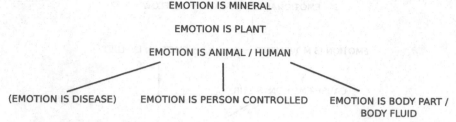

Fig. 2.5 Miscellaneous metaphor themes for emotions

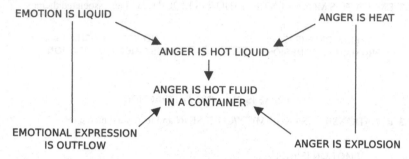

Fig. 2.6 Inter-relations of metaphor themes to produce ANGER IS HOT FLUID IN CONTAINER

might work together to create the much touted ANGER IS HOT FLUID IN A CON-TAINER (for which there is very little lexical evidence), see Fig. 2.6.

2.2.2 Description and Expression of Emotion

Perhaps we should draw a distinction between the metaphorical description and metaphorical expression of emotion. As for the latter, a case can be made for regarding many swear words as metaphorical expressions of emotion. In cases like *piss off*, or *hell*, for example, the mapping or transfer of features is not a matter of conceptual or ideational meaning, but the transfer of the negative feelings about urine or eternal punishment to the meaning of the swear words. These would be clear cut cases of the expression of emotion as one of the interpersonal functions of metaphor [3]. In the metaphorical theme EVIL/WORTHLESSNESS IS WASTE, urine and faeces, 'disgust triggers' [2, p. 174], are used as sources. Faeces metaphorise disgust and contempt for low quality: **shit, turd** 'contemptible, nasty person', **shitty** 'nasty, of low quality', **shit on** 'treat very badly and unkindly', **pooh-pooh** 'show scorn for something' (*he pooh-poohed my attempts to play the piano*). They are also associated with disgust for immorality: **mucky** 'pornographic', **cesspit** or **cesspool** 'unpleasant or immoral situation' (*a cesspit of prostitution and other illegal activities*). Body wastes express contempt for nonsense and uselessness: **crap** 'something useless, worthless, nonsensical or of bad quality' **horseshit** and **bullshit** 'nonsense', **bumf** (literally 'toilet paper') 'written material such as advertisements, or documents that are unwanted or boring'. The same sense of pointlessness or insignificance is found with urine: **piss around** 'waste time doing things without any particular purpose or plan', **piddling** 'insignificant' ($5 *is a piddling amount*).

In most of the above examples metaphorical lexis expresses rather than describes emotion, so it is interpersonal rather than conceptual, but the expression and description of emotion/evaluation often overlap. SAD IS DARK, cited previously as an example of conceptualising emotion, can describe an emotion: *the rise in interest rates* **cast a cloud over** ('induced pessimism about') *the property market*; *news of his father's death* **overshadowed** ('reduced the happiness of') *his winning the gold*. But describing, in the first person, one's own emotion, might amount to expressing it: *this is my* **darkest hour** ('most miserable period of my life') *and I can see no* **light at the end of the tunnel** ('hopes of a pleasant future situation in an unpleasant one'). Either case, expression or description/conceptualisation, involves evaluation.

2.3 Contribution of English Metaphor Themes to the Expression of Emotion

The following discussion of the metaphorical expression of emotion falls into five sub-sections: (1) the problems of mining the Metalude database for emotively expressive metaphors; (2) the question of whether the direction of transfer of emotive evaluation is from source to target or vice versa, or whether there is no transfer; (3) an exploration of the ways in which metaphor themes which are apparently neutral when regarded as stand-alone achieve evaluative force by being part of a larger schema; (4) the dependence of negative or positive evaluation upon ideological stance; and (5) the phenomenon of multivalency and how multivalent sources may converge or diverge in terms of the evaluation of their targets.

2.3.1 Metalude Data for Evaluation

I have sorted through the root analogies/metaphor themes listed in Metalude and attempted to extract those which seem to be evaluative. The rest of this paper is devoted to their discussion. Before I proceed, one caveat. When compiling the lexical data for Metalude, beginning back in the early 90s, my choice of metaphor theme labels was somewhat unsystematic, and heavily reliant on the traditional labels in the CL literature. Nowadays, I would be more systematic, establishing classes and hierarchies according to semantic networks or by exploiting Grady's insights into primary metaphors.

Moreover, I decided, as Metalude is conceived as a resource for teaching Chinese students English vocabulary, to subdivide metaphor themes with more than 50 lexical items. One result is that some metaphor thematic subdivisions draw attention to negative evaluations. For example, as Fig. 2.1 shows, EMOTION IS TOUCH/IMPACT subsumes BAD EMOTION IS HURT/INJURY. However, others,

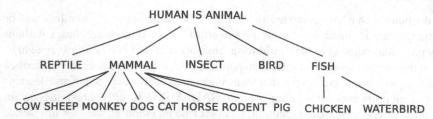

Fig. 2.7 Sub-themes for the metaphor theme HUMAN IS ANIMAL in Metalude

Table 2.1 Metaphor themes from Metalude incorporating evaluative lexis

HUMAN IS ANIMAL (and its subdivisions)	HUMAN IS SUPERNATURAL/MYTHICAL BEING
EMOTION IS WEATHER	MONEY IS FOOD
EXPERIENCE IS FOOD	QUALITY IS MONEY/WEALTH
EXPERIENCE/SITUATION IS WEATHER	QUALITY IS SHAPE/SIZE
HUMAN IS SUPERNATURAL/MYTHICAL BEING	QUALITY IS TASTE/TEXTURE
KNOWLEDGE/WORDS IS FOOD AND DRINK	RANK/VALUE/CHARACTER IS METAL
LANGUAGE QUALITY IS TASTE	WEATHER IS HUMAN ACTIVITY/QUALITY

which are not sub-divided, may also contain a great deal of evaluative lexis, without this being obvious from the metaphor theme label. Clear cases are the metaphor themes with HUMAN IS ANIMAL as their superordinate (see Fig. 2.7). Elsewhere I have shown in detail 'the negative metaphorical slant of these metaphors, many connoting unpleasantness, ugliness, pride, uncontrolled appetite and stupidity' [4, p. 152].

Metaphor themes which have no evaluative term in their title, but which never-theless likely include evaluative terms include those in Table 2.1.

2.3.2 Evaluative Transfer

Some metaphor themes in Metalude have evaluations in both their source and target, for example the negative ones in Table 2.2.

Table 2.2 Metaphor themes from Metalude with negative evaluation in target and source

BAD/UNIMPORTANT IS POOR/CHEAP	PROBLEM/DIFFICULTY IS DISEASE
EVIL/WORTHLESSNESS IS WASTE	AWKWARD SPEECH IS AWKWARD WALKING
BAD IS SMELLY	FAILURE IS SHIPWRECK
EVIL IS DIRT	MENTAL DISTURBANCE IS DIVISION/INCOMPLETENESS
BAD EMOTION IS DISCOMFORT/PAIN	DISEASE IS WAR/INVASION
CAUSE BAD EMOTIONS IS HURT/INJURE	

Table 2.3 Metaphor themes from Metalude with negative transfer from source to target

FEELING EMOTION IS BEING EATEN	
EMOTION/IDEA IS DISEASE	ARGUING/CRITICISING IS
	WOUNDING/CUTTING
CESSATION IS DEATH	ARGUMENTS IS WEAPONS/AMMUNITION
ARGUING/CRITICISING IS ATTACKING	COMPETITION IS WAR/VIOLENCE
ARGUING/CRITICISING IS FIGHTING	SEX IS VIOLENCE
ARGUING/CRITICISING IS	ARGUING/CRITICISING IS
HITTING/PUNCHING	WOUNDING/CUTTING

I have already exemplified WORTHLESSNESS IS WASTE. Another, DISEASE IS WAR/INVASION, shows that some of the lexis under these themes may be positive in its evaluation, despite the pejorative nature of both sides of the label. It constructs disease pejoratively as an **attack** by **invaders** 'viruses or bacteria', or **foreign bodies** from outside. The bacteria **invade** 'enter the body', and may **strike down** 'cause illness or death to' the victims who **succumb** 'become ill'. However, the lexis also contains positive evaluations: the body may **defend** itself, **fight**, **combat** 'struggle to survive' the disease, through **resistance** 'immune response' and medicine might **conquer**, **vanquish** 'eliminate' a disease.

Other themes apparently transfer evaluation from the source label to the target label (Table 2.3), a finding in tune with much of the evidence for conceptual feature mapping of individual lexical metaphors. Some of the lexis for SEX IS VIOLENCE is given in Section 2.3.4 below. As another example, consider EMOTION/IDEA IS DISEASE. Ideas and emotions can be a **bug** 'enthusiasm' that is **contagious**, **catching** or **infectious** 'easily communicated to many people'. (It is worth pointing out however that with these four lexical items the negative evaluation seems to be neutralised by the target, unlike the items below). The emotions associated with ideas can be more or less strong – **virulent** 'full of hate and fierce opposition', **pathological** 'showing extreme uncontrolled feelings'. These ideas and accompanying emotions cause harm – **poison** 'introduce a harmful idea into' the mind or are harmful – **noxious**, **poisonous**, **venomous** 'harmful, negative, unpleasant', **inflammatory** 'intentionally causing negative feelings' or **jaundiced** 'pessimistic', while negative ideas **fester** 'become more intense', like an infected wound.

Resistance is seen in terms of preventing disease: **sanitize** 'change in order to make it less strongly expressed or offensive', **immune** 'unable to be influenced by an idea or emotion'; or of its treatment: **cure of** 'get rid of a bad idea or emotion'.

The majority of evaluative metaphor themes have an evaluative target and an apparently neutral source, but I lack the space to list them all. Table 2.4 lists a sample, some of which I follow up later in the paper.

The politically-incorrect metaphor themes GOOD IS CLEAN/WHITE, and EVIL IS DARK/BLACK are clear examples, there being no intrinsic value to these two colours. Realising the first we have positive lexical items such as **white knight** 'person or organisation that rescues a company from financial difficulties', **fair** 'morally correct or just', **whiter than white** 'having a reputation for high moral

Table 2.4 Metaphor themes in Metalude with an evaluative target

NEGATIVE	POSITIVE
BAD IS LOW	GOOD (MORALITY, QUALITY) IS HIGH
CONFLICTING PURPOSE IS OPPOSITE DIRECTION	SHARE PURPOSE IS ALIGN
UNCERTAINTY/UNRELIABILITY IS INSTABILITY	CERTAINTY/RELIABILITY IS SOLIDITY/FIRMNESS
EVIL IS DARK/BLACK	GOOD IS CLEAN/WHITE
WORRY/PROBLEM/RESPONSIBILITY IS WEIGHT	SERIOUSNESS/IMPORTANCE IS WEIGHT
STEAL IS LIFT	UNDERSTANDING IS PENETRATION/SHARPNESS
DECEIT IS DOUBLENESS	TRUTH/CORRECTNESS IS STRAIGHTNESS

standards', **lily-white** 'faultless in character', and **whitewash** 'cover up mistakes or bad behaviour'. The second comprises mostly pejorative terms, meaning evil or wrong: **black** meaning 'bad' (*this is a black day for the Olympics*), or 'cruel or wicked' (*this is a blacker crime than most I've investigated*), **black and white** 'with clear distinctions between morally wrong and right', **black mark** 'fault or mistake that has been noted', **blackguard** 'a wicked person', **blackleg** 'a traitor who continues to work while other workers are on strike'; they can mean 'illegal', **black market, black economy**; or connote loss of reputation: **black sheep** 'bad person in a family who brings it into disrepute', and **blacken** 'destroy the good reputation of'.

2.3.3 Evaluation Dependent on Larger Schemata

The subdivision into metaphor themes with a more manageable number of lexical items also obscures the fact that many sources are necessarily evaluated if seen as part of a larger schema, though they might not appear to be in isolation.

For instance, one of the more important superordinate schemata for conceptualising activity (life) is movement forwards, ACTIVITY/PROCESS IS MOVEMENT (FORWARD), with its more obviously positive counterparts DEVELOPING/ SUCCEEDING IS MOVING FORWARD and SUCCESS/EASE IS SPEED. As sub-metaphor themes we have some in which neither target nor source are intrinsically evaluative: INACTIVITY IS IMMOBILITY; LESS ACTIVE IS SLOW; and perhaps OPPORTUNITY/POSSIBILITY IS OPENING; PURPOSE IS DIRECTION; PURPOSELESS IS DIRECTIONLESS; SHARE PURPOSE IS ALIGN. Others quite clearly have little evaluation in their sources: CONFLICTING PURPOSE IS OPPOSITE DIRECTION – one might quite happily retrace one's steps after a walk in the country; DIFFICULTY IS MUDDY GROUND – muddy ground is ideal for planting rice; DIFFICULTY/PREVENTION IS OBSTACLE – obstacles are an excellent barrier against threats; FAILURE/GIVING UP IS BACKWARDS – going backwards is the preferred method of parking a car; NO DEVELOPMENT IS IMMOBILITY/CIRCULARITY – round trip holidays always bring you back to where you started, and, indeed taking a circular route around an obstacle is an excellent example of lateral thinking as in SOLUTION IS WAY ROUND/OVER/THROUGH; UNSUCCESSFUL/DIFFICULT IS SLOW – there is nothing intrinsically good about fast and slow (think of drinking good coffee and having good sex). But, as part of the superordinate schema, all these sources are necessarily evaluated negatively as preventing successful activity.

2.3.4 Ideology and Evaluation

This leads us to consider our third question, already explored extensively elsewhere [4]. Evaluation is notoriously variable and subjective, at least relative to conceptual meanings which tend to be more stable within language communities. Ideological divides within society might therefore give us different evaluative stances on the themes in Metalude. This would in some cases amount to a meta-evaluation, not just using evaluative language that represents a counter ideology but responding evaluatively to others' language use. At a deep level, for example, one might observe fundamentally opposed metaphoric models for conceiving humanity and society as in Table 2.5.

The basic distinction is between a structure, in column 1, in which the individual parts are diverse, representing different qualities and therefore incommensurate, and related to each other in a co-operative enterprise. And column 2, where the individual entities are seen as similar and therefore quantifiable, free, separate and in competition with each other. If one espouses a counter-ideology to the current late capitalist one, one might evaluate negatively uses of vocabulary which belongs to the metaphor themes in the second column.

To elaborate and exemplify further, ACTIVITY IS FIGHTING can be divided into sub-themes involving speech acts as in Fig. 2.8. The prevalence of ARGUING/ CRITICISING IS FIGHTING has provoked the following feminist and co-operative response:

> There are non-adversarial aspects of argument. And there are non-adversarial metaphors for argument – arguments may help us **build a case, explore a topic,** or think through a problem. Evaluating arguments may lead us to change our own minds; a critical analysis of someone else's case is not, by definition, a negative one. ([5], my emphasis)

Table 2.5 Ideological and metaphorical oppositions

Relationship	Isolation
Unity	Separation
Diversity	Sameness
Quality	Quantity
Co-operation	Competition
ORGANISATION IS MACHINE	QUALITY IS QUANTITY/SIZE
SOCIAL ORGANISATION IS BUILDING	QUALITY IS WEALTH
SOCIAL ORGANISATION IS BODY	ACTIVITY IS GAME/FIGHTING
RELATIONSHIP IS PROXIMITY/COHESION	FREEDOM IS SPACE TO MOVE

Fig. 2.8 The sub-themes of the metaphor theme ACTIVITY IS FIGHTING

Similarly ecologists (Gaia-theorists) and animal rights campaigners might give a positive evaluation to ANIMAL IS HUMAN, PLACE/LANDSCAPE IS BODY, PLANT IS HUMAN/ANIMAL, as they seem to break down the conceptual barriers between human and non-human nature, and accord a dignity to the latter. Socialists might in principle object to the use of lexis from AFFECTION/RELATIONSHIP IS MONEY/WEALTH and HUMAN IS VALUABLE OBJECT/COMMODITY as they wish to resist the encroachment of the market into every aspect of human life and the commodification of the human body and of relationships. Afro-Caribbeans now prefer to be known as **people of colour**, in order to replace the negative meanings of **black** (Section 2.3.2) with something more exciting (EXCITEMENT IS COLOUR, Fig. 2.1). And feminists would certainly protest against SEX IS VIOLENCE, especially since men are usually constructed as the aggressors: **chopper, weapon, shoot his load, fire blanks, conquest, lady-killer**.

However, as with this last example, careful consideration of the specific lexis realising the metaphor theme is often necessary in order to judge its ideological affinities. Anti-materialists might be thought, for example, to resist HUMAN IS MACHINE/IMPLEMENT. But a close look at its lexis suggests that the number of pejorative lexical items far outweigh the positive (Table 2.6).

Some of this specific pejorative lexis can be interpreted according to the anti-mechanistic dictum that reducing humans to machines or implements demeans them: we are in fact, or should be more than machines. Conservatives' and traditionalists' reactionary hackles would only be raised by careful consideration of the lexis realising UNCHANGING IS HARD/RIGID, and UNCHANGING IS STATIC:

Table 2.6 Pejorative and positive lexis in HUMAN IS MACHINE/IMPLEMENT

PEJORATIVE METAPHORS	POSITIVE METAPHORS
Crook, crock, rake, basket-case, hatchet-faced, flail, rasping; mechanically, automaton, motormouth, cog, crank, hulk	New-broom, dynamo, drive, turbocharged, high-powered, high-octane

UNCHANGING IS HARD/RIGID

NEGATIVE **unyielding, stiff-necked, hard-line**, unwilling, stubborn or unable to change their beliefs or behaviour; **rigid, rigidity**, obstinately resisting change or persuasion; **starchy**, old-fashioned and formal in behaviour; **fossilized, petrified, ossified**, unable to change or develop positively; **set**, unchanging, conservative; **embedded**, unchanging, permanent; **unbending**, tending to make judgments that cannot be changed; **set/cast in stone/concrete**, extremely difficult to change

POSITIVE: **stable**, not likely to change for the worse; **solid**, certain, unwavering, loyal

UNCHANGING IS STATIC

NEGATIVE **stuffy**, formal, boring and old-fashioned; **stick in the mud**, someone who is not willing to change or accept new ideas; **cling to**, refuse to give up a tradition or belief; **entrenched**, difficult or impossible to change **dig in/dig their heels in**, refuse to change your opinions or plans, **stuck with, tied to**, forced to accept a situation you cannot change, **stagnant, stagnate** fail/-ing to change develop or improve

POSITIVE: **settled**, permanent and predictable, **stick at, apply yourself to**, keep doing the same thing with determination despite difficulties, **stick by**, continue to give help and support to a person

In a similar way, there is nothing which alerts us to sexist ideology in the label HUMAN IS FOOD. However, the lexis indicates women are disproportionately represented as food, where their purpose is to satisfy the appetites of men: **cheesecake** 'half-naked, female, photographic models', **crackling, crumpet** 'sexually attractive woman', **tart** 'sexually immoral/attractive woman', **mutton dressed as lamb** 'older woman trying to look young', **lollipop, peach** 'attractive young girl', **arm-candy** 'attractive companion at social events'. (Though we also have **dishy, stud muffin**, and **beefcake** applied exclusively to men).

2.3.5 The Role of Multivalency and Opposition in Metaphor Themes

As suggested elsewhere, metaphor themes can interact in interesting ways to affect our cognition and ideological value judgments, for example attitudes to immigration and race [4, Chapter 5]. What interests me in this section is the way in which multivalent metaphor themes, those with an identical source and different targets, may converge (or diverge) in terms of negative and positive evaluation. For example the four themes with straightness as source all seem positive.

GOODNESS (HONESTY) IS STRAIGHTNESS

JUSTICE/LAW IS STRAIGHT (LINE)

TRUTH/CORRECTNESS IS STRAIGHTNESS

SANITY/*NORMALITY IS STRAIGHTNESS

Table 2.7 Metaphor themes from Metalude involving 'height' as source

POSITIVE	NEGATIVE
HIGH	**LOW**
GOOD(QUALITY/MORALITY) IS HIGH	BAD IS LOW
HAPPY IS HIGH	SAD IS LOW
HEALTH/LIFE IS HIGH	UNHEALTHY/DEAD IS LOW
POWER/CONTROL IS ABOVE	POWERLESS/CONTROLLED IS BELOW
IMPORTANCE/STATUS IS HIGH	UNIMPORTANT/SUBORDINATE IS LOW
*MORE IS HIGH	*LESS IS LOW
(CAUSE) TO GO/BE HIGH	**(CAUSE) TO GO/BE LOW**
BE GOOD ENOUGH/BETTER IS RISE	DETERIORATE IS FALL/LOWER
GAIN POWER IS RISE	LOSE POWER/CONTROL IS DESCEND
ENCOURAGE/HELP IS SUPPORT	CONTROL IS PUSH/PUT DOWN
IMPROVE STATUS IS RAISE	REDUCE STATUS IS LOWER
ACHIEVEMENT/SUCCESS IS HIGH	FAILURE IS FALLING
	FAILURE IS SINKING
*INCREASE IS RISE	*DECREASE IS FALL

Probably the most obvious pattern of significant multivalency concerns the source of height (Table 2.7).

In cases of both STRAIGHTNESS and HEIGHT I have placed asterisks against targets which do not, at face value, seem positive or negative. Normality may be boring, having more work to do may be negative. I suggest that the sharing of sources may bring about a sharing of evaluative polarity: if GOOD IS HIGH and MORE IS HIGH, then MORE = GOOD [4, Chapter 5]. This becomes even more pronounced with the metaphor themes with the multivalent source BIG.

> IMPORTANT IS BIG
> *NUMEROUS/MORE IS BIG
> *FEW/LESS IS SMALL
> *INCREASE IS EXPAND
> *DECREASE IS CONTRACT

Most of these are asterisked, and it would seem that only under the influence of IMPORTANT IS BIG do they achieve a positive evaluation for large size and negative for small size. Notice, therefore, that in none of the metaphor themes cited in this section does the source intrinsically carry an evaluation, and that evaluation is either achieved by transfer from the target, or from other targets which share the same source.

However there are some conflicting evaluations, for example WEIGHT can be given a positive evaluation as in SERIOUSNESS/IMPORTANCE IS WEIGHT or a negative one as in WORRY/PROBLEM/RESPONSIBILITY IS WEIGHT. This difference depends upon the primary scenes or schemata and metonymic frames to which weight belongs in each case. Seriousness and importance might be associated with the weighing of goods, coins or metals, in which schema it is positive, while worry, problems or onerous responsibilities might be associated with DEVELOP-MENT/SUCCESS IS MOVEMENT FORWARDS (LIFE IS A JOURNEY), or SAD

IS LOW (manifested by slouching gait, drooping shoulders downcast eyes), in which case the weight is a burden and impediment to movement or upright posture.

An important conflicting evaluation of a similar source arises with RELATION-SHIP IS PROXIMITY/COHESION and NO FREEDOM IS TYING/BINDING, one of the converses of FREEDOM IS SPACE TO MOVE (Table 2.5). This conflict tends to construct relationships as a loss of freedom, rather than the means of achieving identities and roles through which we are empowered, and in which 'service is perfect freedom' [4].

2.4 Conclusion

Cognitive linguistics, as the label suggests, has for the most part concentrated on the conceptual or ideational aspects of meaning, and hence has had a great deal to say about the conceptualisation of emotion. It has, however, more or less neglected the interpersonal aspects of metaphor use, of which the expression of emotion is one. Though the lexical resources for conceptualisation/description and expression in some cases overlap, as when 1st person description amounts to expression, in other cases, such as swear words, expression is quite distinct from conceptual meaning and depends on affective grounds. So, while the first part of this paper is treading on well-worn ground, albeit beating a lexicological rather than an intuitive path, the second part is more exploratory, and I hope, opens the way for more research. I have tried to show that data may be mined from Metalude not only to reveal how emotion is conceptualised or described but, somewhat problematically, to uncover the metaphorical evaluative resources in English.

References

1. Deignan, A. 2005. *Metaphor and corpus linguistics*. Amsterdam: Benjamins.
2. Eckman, P. 2000. *Emotions revealed*. London: Weidenfeld and Nicholson.
3. Goatly, A. 1997. *The language of metaphors*. London and New York: Routledge.
4. Goatly, A. 2007. *Washing the brain: Metaphor and hidden ideology*. Amsterdam: Benjamins.
5. Govier, T. 1999. *The philosophy of argument*. Newark Port, VA: Vale Press.
6. Lakoff, G. and M. Johnson. 1980. *Metaphors we live by*. Chicago: University of Chicago Press.
7. Metalude. 2005. http://www.ln.edu.hk/lle/cwd03/lnproject_chi/home.html. User id: (user), password: (edumet6).

Chapter 3
The Deep Lexical Semantics of Emotions

Jerry R. Hobbs and Andrew S. Gordon

3.1 Introduction

We understand discourse so well because we know so much. If we are to have natural language understanding systems that are able to deal with texts with emotional content, we must encode knowledge of human emotions for use in the systems. In particular, we must equip the system with a formal version of people's implicit theory of how emotions mediate between what they experience and what they do, and rules that link the theory with words and phrases in the emotional lexicon.

The effort we describe here is part of a larger project in knowledge-based natural language understanding to construct a collection of abstract and concrete core formal theories of fundamental phenomena, geared to language, and to define or at least characterize the most common words in English in terms of these theories [8]. One collection of theories we have put a considerable amount of work into is a commonsense theory of human cognition, or how people think they think [9]. A formal theory of emotions is an important piece of this. In this paper we describe this theory and our efforts to define a number of the most common words about emotions in terms of this and other theories.

Vocabulary related to emotions has been studied extensively within the field of linguistics, with particular attention to cross-cultural differences [1, 6, 18]. Within computational linguistics, there has been recent interest in creating large-scale text corpora where expressions of emotion and other private states are annotated [17].

In Section 3.2 we describe Core WordNet and our categorization of it to determine the most frequent words about cognition and emotion. In Section 3.3 we describe an effort to flesh out the emotional lexicon by searching a large corpus for emotional terms, so we can have some assurance of high coverage in both the core theory and the lexical items linked to it. In Section 3.4 we sketch the principal facets of some of the core theories. In Section 3.5 we describe the theory of Emotion with several examples of words characterized in terms of the theories.

J.R. Hobbs (✉)
Information Sciences Institute, University of Southern California, Marina del Rey, CA 90292, USA
e-mail: hobbs@isi.edu

K. Ahmad (ed.), *Affective Computing and Sentiment Analysis*, Text, Speech
and Language Technology 45, DOI 10.1007/978-94-007-1757-2_3,
© Springer Science+Business Media B.V. 2011

3.2 Identifying the Core Emotion Words

WordNet [14, 15] contains tens of thousands of synsets referring to highly specific animals, plants, chemical compounds, French mathematicians, and so on. Most of these are rarely relevant to any particular natural language understanding application. To focus on the more central words in English, the Princeton WordNet group has compiled a CoreWordNet, consisting of 4,979 synsets that express frequent and salient concepts. These were selected as follows: First, a list with the most frequent strings from the British National Corpus was automatically compiled and all WordNet synsets for these strings were pulled out. Second, two raters determined which of the senses of these strings expressed "salient" concepts [2]. CoreWordNet is downloadable from http://wordnet.cs.princeton.edu/downloads.html.

Only nouns, verbs and adjectives were identified in that effort, but subsequently 322 adverbs were added to the list.

We classified these word senses manually into sixteen broad categories, including such classes as Composite Entities, Scales, Events, Space, Time, Communication, Microsocial (e.g., personal relationships), Macrosocial (e.g., government), Artifacts, and Economics. A very important class was Cognition, or concepts involving mental and emotional states. This included such words as *imagination, horror, rely, remind, matter, estimate*, and *idea*. Altogether 778 words senses were put into this class.

These were further divided into thirty classes based on commonsense theories of cognition we had identified from an examination of several hundred human strategies [4] and had constructed formal theories of in a defeasible, first-order predicate calculus [9]. Among the thirty are theories of Knowledge Management, Memory, Goals and Plans, Envisionment (or "thinking about"), Decisions, Threat Detection, Explanations, and Emotions. 140 of the 778 cognitive word senses concern emotions, and are the focus of this paper. Some random examples of the emotion word senses are as follows (many of these are ambiguous, but it is the emotional sense that concerns us): *heart, concern, relief, anger, mood, joy, fit, embarrassment, morale, apathy, pride, disgust, want, feel, suffer, cry, upset, provoke, terrify, fascinate, glad, exciting, happy, sympathetic, passionate*, and *calmly*.

3.3 Filling Out the Lexicon of Emotion

With the aim of providing automated tools for annotating expressions of emotion in English text, we developed a catalogue of English words and phrases that refer to emotional states and emotion-related mental events, as part of a larger effort to recognize all English expressions related to commonsense psychology [5].

Our strategy consisted of three steps. First, we convened a group brainstorming meeting with researchers, graduate students, and administrative staff within our research lab. Participants were asked to creatively and competitively produce words and phrases that were related to emotional states, the expression of emotions, and commonsense mental processes involving emotions. The purpose of this meeting was to produce an initial list that could serve as the starting point for an exhaustive linguistic search. Second, a team of graduate students in linguistics and computational linguistics were tasked to elaborate this list by consulting a variety of thesauri, phrase dictionaries, and electronic linguistic resources. WordNet was particularly useful during this step; the list was expanded to include all hyponyms of *emotion*-1, troponyms of *provoke*-1, and troponyms of *feel*-1. Morphological derivatives of each word in the expanded list were also included, e.g., the verb *resent* relates both to its present participle (*resenting*), but also to the adjective *resentful* and its derivatives (*resentfully* and *resentment*). Third, the resulting list (several hundred emotion terms) was then organized into semantic classes by clustering terms with similar meaning. During this step, we relied heavily on the emotion categories proposed by Ortony et al. [16], expanded by Clark Elliott [3] to include 24 distinct emotion types. The final taxonomy added a superordinate emotion class, a class for the lack of emotion, and seven classes of terms related to emotion-related mental processes, resulting in a final list of 33 taxonomic distinctions.

In conducting this analysis, we were particularly struck by two characteristics of emotion vocabulary that distinguishes it from other terminology related to commonsense psychology, e.g. beliefs, goals and plans. First is the sheer quantity of single words that reference emotion states in the English language, in no small part due to the borrowing power of English; there are literally hundreds of words available to English-speakers to describe how they are feeling. Second is the low level of polysemy within this set; most emotion terms have only a single word sense. The list below provides several examples of each of the 33 emotion categories, with the adjectival form favored over other derivatives.

1. emotion (*affect, emotion, feeling, have feelings of*)
2. joy emotion (*blithe, cheery, comfortable, ecstatic, elated, enjoyment, happy, be in high spirits, be in Nirvana, be on cloud nine*)
3. distress emotion (*agony, bereavement, brokenhearted, cheerless, depression, despondent, sad, tearful, unhappy, be low spirited, have a sinking feeling*)
4. happy-for emotion (*glad for, pleased for, congratulatory*)
5. sorry-for emotion (*commiserative, compassionate, condolence*)
6. resentment emotion (*covetous, envious, jealous, sulky, vengeful*)
7. gloating emotion (*schadenfreude, mawkish*)
8. hope emotion (*encouragement, hopeful, optimistic, sanguine*)
9. fear emotion (*anxious, apprehensive, bode, consternation, despair, fearful, terror, timid, trepidation, uneasy, worried, have cold feet, gives one the creeps*)
10. satisfaction emotion (*consolation, delightful, gratification, pleasure, ravishment, satisfaction, solace, have a silver lining*)
11. fears confirmed emotion (*fears have come true, fears realized*)
12. relief emotion (*alleviation, assuagement, relief*)
13. disappointment emotion (*defeat, disappointment, frustration*)
14. pride emotion (*conceited, egotistic, proud, prideful, vain*)
15. self-reproach emotion (*chagrin, discomfit, embarrassment, humble, humility, meek, repentance, self-conscious, self-depreciation, shame*)

16. appreciation emotion (*appreciative, thankful*)
17. reproach emotion (*disapproval, reproachful*)
18. gratitude emotion (*grateful*)
19. anger emotion (*aggravation, angry, annoyance, belligerent, furious, pique, rage*)
20. gratification emotion (*gratifying*)
21. remorse emotion (*guilt, regretful, remorseful, rueful*)
22. liking emotion (*fancy, fascination, fondness, partiality, penchant, predilection, have a taste for, have a weakness for*)
23. disliking emotion (*abhorrent, abomination, detestable, disinclination, dislikable, execration, loathsome, repugnant, repulsive, revulsion*)
24. love emotion (*adoration, agape, amorous, devotion, enamor, infatuation, lovable*)
25. hate emotion (*animosity, bitterness, despise, hateful, malefic, malevolent, malicious, spite, venomous, have bad blood*)
26. emotional state (*mood, way one feels, how one is feeling*)
27. emotional state explanation (*reason for feeling, why one feels, cause of the emotion*)
28. emotional state change (*a shift in mood*)
29. appraisal (*assess one's emotions, figure out how one feels about*)
30. coping strategy (*way of dealing with, coping technique*)
31. coping (*dealing with the feeling, coming to terms with*)
32. emotional tendency (*emotional, moodiness, passionate, sentimentality*)
33. no emotion (*aloof, ambivalent, austere, calm, cold-hearted, emotionless, heartless, impassive, indifferent, phlegmatic*)

3.4 Some Core Theories

We use first-order logic for encoding axioms in our commonsense theories, in the syntax of Common Logic [13]. Since human cognition concerns itself with actual and possible events and states, which we refer to as eventualities, we reify these and treat them in the logic as ordinary individuals. Similarly, we treat sets as ordinary individuals and axiomatize naive set theory. Most axioms are only normally true, and we thus have an approach to defeasibility – proofs can be defeated by better proofs. Our approach to defeasibility is based on weighted abduction [11] and is similar to McCarthy's circumscription [12], but the content of the theories should survive a translation to any other adequate framework for defeasibility.

The theories of cognition rest on sixteen background theories. Included among these is a theory of scales that provides means of talking about partial orderings, the figure-ground relation of placing some external thing *at* a point on a scale, and qualitative regions identifying the *high* and *low* regions of a scale. The latter are linked to the theory of functionality mentioned below; often when we call something tall, we mean tall enough for some purpose. They also need to be linked to an as-yet undeveloped commonsense theory of distributions.

In addition, we have theories of change of state, causality, and time. The theory of causality tries to provide a defeasible notion of *cause* that can be used in lexical semantics [7]. The theory of time explicates such predicates as *before, atTime* relating an event to a time, and a *meets* relation between intervals [10].

For this paper the most relevant cognitive theories are Knowledge Management, Goals and Planning, and Envisionment. In the theory of Knowledge Management, we characterize belief and graded belief and their relation to perception, inference, and action. Briefly, perceiving is believing, we can defeasibly do logic inside belief contexts, and our beliefs influence our actions. We also axiomatize change of belief, mutual belief, assuming, varieties of inference, justification, knowledge domains, expertise, and other similar concepts in this theory.

The theory of Goals and Planning posits agents that have a top-level goal "to thrive", have various beliefs about what will cause them to thrive and other causal knowledge, and continually plan and replan to achieve this top-level goal. Planning uses axioms about what eventualities cause or enable what other eventualities to generate subgoals of goals, and subgoals of the subgoals, until arriving at executable actions. Shared goals and plans are defined in terms of mutual knowledge and of sets of agents having goals where the shared plans bottom out in actions by individual members. We define notions of eventualities being good for or bad for an agent or group of agents relative to their goals. The function and roles of artifacts and organizations are characterized in terms of agents' goals, where the structure of the artifact or organization reflects the structure of the plan to achieve the goals. We also explicate here the notions of attempting to achieve a goal and actually achieving it. A *threat* is an eventuality that may cause one's goals not to be achieved.

The theory of Envisionment is an attempt to begin to capture what it is to think about something, particularly, in a causal manner. To envision is to entertain in one's focus of attention a sequence of causally linked sets of eventualities. For example, the Common Logic expression (`envisionFromTo a s1 s2`) says that an agent a envisions a sequence of causally connected situations starting with s1 and ending with s2. Explanation, prediction, and planning are varieties of envisionment.

3.5 The Theory and Lexical Semantics of Emotion

Our theory of Emotions attempts to characterize twenty-six basic emotions in terms of the abstract situations that cause them and the abstract classes of behavior they trigger. That is, emotions are viewed primarily as mediating between perception and action. Our treatment is based in part, but only in part, on that of Ortony et al. [16]. We attempt, in addition, to axiomatize the notion of the *intensity* of emotion, and give a somewhat more central role to the "raw emotions", as described below.

Natural language is very rich in emotional terminology, and our formal theory of emotion tracks language very closely. Thus, in explicating the concepts of the theory, we are also providing the deep lexical semantics of English emotional terms. Of course, the converse is not also true; there are many more English emotional

terms than would be basic predicates in an underlying theory of emotion; these
others we characterize in terms of the basic predicates.

Happiness is normally caused by the belief that one's goals are being satisfied.
This of course is not always the explanation of one's happiness. Imagining you will
win the lottery can cheer you up, sometimes you feel happy for no identifiable reason
at all, and sometimes you are unhappy even though everything is going well. This is
an illustration of why virtually all the rules in the cognitive theories are defeasible.

To give a flavor of the rules in the theories, we include the fairly complex one
characterizing one of the sources of happiness.

```
(forall (a g e1 e2 e3 t1 t2)
   (if (and (goal' e1 g a)
            (atTime e1 t1)
            (atTime' e2 g t2)
            (believe' e3 a e2)
            (atTime e3 t1)
            (intMeets t1 t2) <etc>)
        (exists (e4)
           (and (happy' e4 a)
                (atTime e4 t1)
                (cause e3 e4))))))
```

That is, if during time interval t1 agent a has the goal g and believes that it will
be satisfied during interval t2, where t2 begins when t1 ends, then this belief will
cause a to be happy during interval t1. More succinctly, anticipating success makes
us happy. The <etc> is an abbreviation indicating defeasibility.

An inference one can draw from one's success in satisfying one's goals is that
the rules or beliefs that generate one's behavior are functional. They are the right
rules. Therefore, there are two conclusions with respect to one's actions. Since the
rules are correct, there will be a reluctance to change one's beliefs, at least in the
relevant knowledge domains. The current beliefs are doing a good job. And one will
be inclined to act on one's current beliefs. One will exhibit a greater level of activity.

Sadness is given a corresponding characterization. It is normally caused by the
belief that one's goals are not being satisfied. It tends to suppress the urge to action,
since one would be acting on beliefs that have shown themselves to be dysfunc-
tional. Moreover, sadness opens one to a change in beliefs.

We have axiomatized Ortony et al.'s [16] cognitive elaborations on basic emotions. Happiness and sorrow for someone else, resentment, and gloating are defined
in terms of eventualities being good for or bad for in-groups and out-groups, where
in-groups are defined in terms of shared goals. Anticipation is defined in terms of
envisionment; satisfaction, "fears confirmed", disappointment, and relief are defined
in terms of anticipated eventualites that are good for or bad for the agent being
realized or frustrated. Pride, self-reproach, appreciation, reproach, gratification,
remorse, gratitude and a certain kind of anger are defined in terms of eventualities
that are good for or bad for one's self or others being merely attempted or
succeeding.

Although we do not "define" emotional intensity, we do constrain its interpretations with axioms that say in some special circumstances what sort of emotions will normally be more intense than others, *ceteris paribus*. For example, normally the more salient the stimulus, the more intense the emotion, and the more intense the emotion the more extreme the response. *Intense* then labels the functionally and distributionally high region of that scale.

Our treatment of the three "raw" emotions, anger, fear, and disgust, depends on the notions of eliminating or avoiding threats. One eliminates a threat by causing a change of state (or location) in it. One avoids a threat by causing a change of state (or location) in one's self. In either case, the effect is a reduction of the threat. Anger and fear are both caused by threats. In anger, our response to it is normally to try to eliminate the threat. In fear, our response is normally to try to avoid the threat.

Fear and anger are responses to external threats. Disgust is a response to a threat that is interior, and it triggers an effort to eject the threat. "Interior" may be interpreted literally with respect to the body – most of the ways of talking about disgust involve distaste or nausea. Or we may interpret it metaphorically as referring to an in-group.

All of this is of course quite naive if viewed as a *real* theory of emotions. But we believe it is reasonable as a commonsense theory, and will allow natural language systems to make sense of most occurrences of emotion terms in English discourse.

Having explicated the basic emotions formally, we are now able to write axioms characterizing the meanings of the less central emotional terminology of English. For example, to "terrify" someone is to cause one to feel intense fear. The various emotional word senses of "calm" in WordNet can be characterized in terms of feeling or causing low emotional intensity.

There are five noun senses of *pride* in WordNet. *pride-N2* includes the Ortony et al.'s [16] sense we characterized above as what one feels on an attempt to do something good, but also includes the feeling on success and the feeling about another person's attempt or success. *pride-N1* is a version of *pride-N2*, generalized over time. *pride-N3* refers to the causal power of *pride-N1* in one's actions. *pride-N5* is *pride-N1* carried to excess. (The fourth sense is a group of lions.) The single verb sense of *pride* means to feel or express *pride-N1*.

3.6 Summary

Natural language understanding requires a large knowledge base of commonsense knowledge that explicates concepts in coherent theories and links lexical items with these theories. In order to achieve high accuracy, high complexity results, this effort must be manual (as indeed dictionaries are constructed manually). Early efforts will have the most impact if done for the most central concepts and the most common word senses.

In this paper we have outlined our work in constructing background theories and theories of general cognition, and we have described in more detail the structure of the theory of Emotion, indicating how it can be used to explicate the emotional vocabulary of English.

References

1. Athanasiadou, Angeliki and Elzbieta Tabakowska. eds. 1998. *Speaking of emotions: Conceptualisation and expression.* Berlin, NY: Mouton de Gruyter.
2. Boyd-Graber, Jordan, Christiane Fellbaum, Dan Osherson, and Robert Schapire. 2006. Adding dense, weighted, connections to WordNet. Proceedings, Third Global WordNet Meeting, Jeju Island, Korea, January 2006.
3. Elliott, Clark. 1992. The affective reasoner: A process model of emotions in a multi-agent system. PhD diss., Northwestern University. The Institute for the Learning Sciences, Technical Report No. 32.
4. Gordon, Andrew S. 2004. *Strategy representation: An analysis of planning knowledge.* Mahwah, NJ: Lawrence Erlbaum Associates.
5. Gordon, Andrew S., Abe Kazemzadeh, Anish Nair, and Milena Petrova. 2003. Recognizing expressions of commonsense psychology in English text. Proceedings, 41st annual meeting of the Association for Computational Linguistics (ACL-2003), Sapporo, Japan, July 2003.
6. Harkins, Jean, and Anna Wierzbicka, eds. 2001. *Emotions in crosslinguistic Perspective.* Berlin, NY: Mouton de Gruyter.
7. Hobbs, Jerry R. 2005. Toward a useful notion of causality for lexical semantics. *Journal of Semantics* 22:181–209.
8. Hobbs, Jerry R. 2008. Deep lexical semantics. Proceedings, 9th international conference on Intelligent Text Processing and Computational Linguistics, Haifa, Israel, February 2008.
9. Hobbs, Jerry R., and Andrew S. Gordon. 2005. Encoding knowledge of commonsense psychology. Proceedings, 7th international symposium on Logical Formalizations of Commonsense Reasoning, Corfu, Greece, pp. 107–114, May 2005.
10. Hobbs, Jerry R. and Feng Pan. 2004. An ontology of time for the Semantic Web. *ACM Transactions on Asian Language Information Processing* 3:66–85.
11. Hobbs, Jerry R., Mark Stickel, Douglas Appelt, and Paul Martin. 1993. Interpretation as abduction. *Artificial Intelligence* 63:69–142.
12. McCarthy, John. 1980. Circumscription: A form of nonmonotonic reasoning. *Artificial Intelligence* 13:27–39.
13. Menzel, Chris, et al. 2008. Common logic standard. http://cl.tamu.edu/.
14. Miller, George. 1995. WordNet: A lexical database for English. *Communications of the ACM* 38:39–41.
15. Miller, George, Christiane Fellbaum, and Randee Tengi. 2006. WordNet: A lexical database for the English language. http://wordnet.princeton.edu/.
16. Ortony, Andrew, Gerald L. Clore, and Allan Collins. 1988. *The cognitive structure of emotions.* New York: Cambridge University Press.
17. Wiebe, Janyce, Theresa Wilson, and Claire Cardie. 2005. Annotating expressions of opinions and emotions in language. *Language Resources and Evaluation* 39:165–210.
18. Wierzbicka, Anna. 1999. *Emotions across languages and cultures: Diversity and universals.* Cambridge: Cambridge University Press.

Chapter 4
Genericity and Metaphoricity Both Involve Sense Modulation

Carl Vogel

4.1 Background

I wish to explore the link between interpretation of metaphors and generics in natural language, in support of a claim that the mechanisms and processes of interpretation for metaphors and generics are closely related through word sense modulation. Both tropes have curious truth conditions. In a strict literal sense (4.1), (4.2), (4.3), and (4.4) are false.

Sumo wrestlers are elephants.	(4.1)
Sumo wrestlers are bean-poles.	(4.2)
Sumo wrestlers are Japanese.	(4.3)
Sumo wrestlers are Dutch.	(4.4)

Strict literal senses depend upon universal applicability to individuals of the kinds about which the predications are made. No sumo wrestler really is an elephant, and there are many counterexamples to any claim that all sumo wrestlers are Japanese, such as the reading of (4.3) with implicit universal quantification suggests. Loose literal senses depend on existential assertions about the applicability to some individual or other as a member of a "witness set" in support of the claim.[1] A loose literal sense may be regarded as non-literal. It is reasonable to assert, in a non-literal sense for each, that both (4.1) and (4.3) are true (or to deny them).[2] The example (4.3), with a bare-plural subject, can be used to express either that all sumo wrestlers are Japanese ("strict", but false) or that some are ("loose", and true). In the strict literal sense, non-negated metaphors and generics are false; however, it is loose evaluation that appears to underpin common use of both. I argue that both metaphors like (4.1)

---

[1] Witness sets, as invoked in generalized quantifier theory, explain how the cognitive load required to evaluate predications of noun phrases depends on the determiners' monotonicity properties [2].
[2] For an example of (4.3) used as a generic, see: http://answers.yahoo.com/question/index?qid=20080320042727AALZv3Z – last verified January 2011.

C. Vogel (✉)
Computational Linguistics Group, Intelligent Systems Laboratory, Department of Computer Science, O'Reilly Institute, Trinity College (University of Dublin), Dublin 2, Ireland
e-mail: vogel@cs.tcd.ie

K. Ahmad (ed.), *Affective Computing and Sentiment Analysis*, Text, Speech and Language Technology 45, DOI 10.1007/978-94-007-1757-2_4,
© Springer Science+Business Media B.V. 2011

and generics like (4.3) can be understood in terms of belief revision in first order languages augmented with sense distinctions. In this framework, both metaphors and generics are contingent (e.g. (4.2) and (4.4) are false in their respective special senses).

Negations highlight the contingency of metaphors and generics further. The canonical example of negated metaphor, Donne's (4.5), can be used to show that the negation of a metaphor is "patently" true [7]. Less aphoristic examples clarify that the truth of negated expressions, like non-negated ones, depends on the situations described. Examples (4.6) and (4.7) contain sentential negation. These are strictly true. They can also be seen as metaphorically false (if evaluated in situations that contain individuals who are extremely massive in relation to normal body mass for sumo wrestlers). Moreover, the form of negation interacts: (4.8), which involves a negative determiner in the subject noun phrase, is also strictly true. However, (4.7) can be metaphorically true in situations where (4.8) is metaphorically false, such as those where some sumo wrestlers are aptly characterized as elephants and some are not.

No man is an island.   (4.5)
It is not the case that sumo wrestlers are elephants.   (4.6)
Sumo wrestlers are not elephants.   (4.7)
No sumo wrestler is an elephant.   (4.8)
It is not the case that sumo wrestlers are Japanese.   (4.9)
Sumo wrestlers are not Japanese.   (4.10)
No sumo wrestler is Japanese.   (4.11)

Where metaphoricity of the predication is not at stake, but rather the genericity of the utterance, under a strict literal interpretation as above, the sentential negation makes (4.9) and (4.10) true, since it is not the case that all sumo wrestlers are Japanese. In fact, this strict reading of the bare plural subject as involving universal quantification within the scope of the negation seems strongly dis-preferred. Allowing a loose, generic reading makes the truth depend on regularities in the world (in which case, it is false if focus is restricted to the Japanese wrestlers, and true if focus includes the sumo wrestlers born outside Japan). Interestingly, the negative determiner blocks a generic reading for (4.11), but in any case the truth of falsity of the sentence depends on facts about the world and with which sense one wishes to evaluate the sentences. I am concerned here with both the contingency of metaphorical and generic assertions and the constraints on interpretation introduced by negation.

Influenced by work in dynamic semantics that formalized accounts of anaphora in discourse as eliminating possible models of sentences with pronouns, on the basis of restricting assignment functions that map variables into the domain, as pronouns are resolved to potential antecedents [13, 16], as well as research in belief revision [1, 22] Oliver Lemon proposed a framework for first-order logical languages which admitted both information increase and retraction ("updates" and "down-dates", respectively). Carl Vogel [27] proposed a comparable system for information increase only, but with the additional dimension of intensionality in that indices for interpretation were provided to account for the multiplicity of senses that a predicate name or name of individuals might have. That system provided for classical static interpretation (but relativized to senses) and dynamic interpretation, which in all but certain well-defined syntactic and semantic contexts may allow the update and downdate of characteristic functions of sets that provide denotations of relation names and constants. Metaphoricity is captured as a partial order that classifies indices, thus accommodating the intuition that today's novel metaphor is tomorrow's conventionalized non-literal expression, and the next day's dead metaphor, literal language. The system exploits the fact that natural languages supply mechanisms to indicate that non-literal interpretation is intended. For example, it has been noted that the appearance of "literally" in a sentence is a fairly reliable indicator that the sentence it appear in is not to be interpreted literally [12]. It also exploits languages' internal means of disambiguating the intended sense of an expression (even if these are periphrastic, for example, "I mean 'bank' in the sense of 'a financial institution' "). The framework offers a proof-of-concept response to Davidson's claim that metaphor is not within the remit of semantics, but of pragmatics [7]. Carl Vogel [27] provided a truth-functional compositional semantics that could accommodate metaphor and sense extension (expansion of predicates to new entities, and multiple senses for names of entities and relations), but rejected Davidson's claim that "special senses" are not involved in metaphoricity.

In contrast, it has been argued that natural language generics, phenomena well studied in the formal semantics of natural language [3–5, 15, 19], are not in the remit of semantics but of mathematical formulation of a cognitive theory of concepts [29]. One claim made to support this argument is that unlike the case of metaphor, there are no overt markers of genericity. While there is ample treatment of the ability of definite NPs, bare plurals, mass nouns and even indefinite singulars to sustain generic readings, they do not demand them. This ignores the possibility that the unmarked case is generic reference, such as in *determinerless* classifier languages where the specific reading is optionally marked as such if context does not clarify.

Hurricanes happen in the Atlantic and Caribbean.	(4.12)
Leslie smoked cigarettes.	(4.13)
Leslie smoked three cigarettes.	(4.14)

Habituals (4.12) with unbounded subjects, and comparable constructions with terminative aspect (see [29]) make this more clear: without a specific bound or clear definite marking on the object NP in (4.13), the preference is to understand the sentence as a past tense habitual, a form of generic. On the other hand, (4.14) exhibits terminative aspect. The test between the two potential readings is in whether the sentence tolerates modification by "for a day" or "in a day" – (4.13) can be continued with "for a day" but not "in a day", and (4.14) has the reverse pattern. To obtain the specific episodic reading, explicit marking is necessary on the object NP.[3]

---

[3] Sheila Glasbey [9] notes that aspectual class can diverge between literal and non-literal readings of idiomatic expressions.

This article argues sense modulation processes are shared by metaphoricity and genericity. The theory invokes first-order languages which include sense-selection, traditional static interpretation and dynamic interpretation [28, 29].[4] The theory discriminates between the interpretation requirements of novel and established metaphors. The same framework is used to model aspects of both metaphoricity and genericity (the former is expansive, and the latter is restrictive in subsequent interpretation potential). This analysis resonates with one dominant theory of metaphor understanding that holds metaphors to be class inclusion statements [9, 10, (Chapter 1 by Sam Glucksberg, 2011, this volume)]. Thus, the paper also argues that the semantic analysis advocated here is compatible with and extends important aspects of Glucksberg's theory for nominal metaphors.

Section 4.2 characterizes a formal system for update and downdate [29] which is slightly richer than the starting point provided by [22] (it does not require that every element in the domain have a name; it admits multiplicity of sense; it admits sense designation into the language) and is conceptually more complete than the framework provided by [28] in forcing a clear separation between information assertion and retraction and the role of metaphoricity (Section 4.3). Section 4.4 demonstrates how the resulting system provides the restricted quantification of genericity (generics are also analyzed with special non-literal senses). Finally, the paper shows how some of the desiderata of Glucksberg's theory are met. The main explanatory mechanism of Glucksberg's theory is allowance of dual reference in the vehicle of a metaphor in its predication of the topic, ambiguous in predication of the topic between literal reference and an abstraction over that reference that retains salient attributable properties. Asymmetries of metaphors (in contrast to the symmetry of similes) are anchored in the distinction between given and new information, with respect to qualifiable dimensions in the given information and potential attributions supplied by the new information. Other desiderata (for example, conflation of subject-object asymmetry in metaphors with topic-comment information packaging) are disputed.

4.2 Dynamics of First-Order Information

4.2.1 Some Intuitions About Revision

To a child learning about the world from documentaries, it may be news that (4.15) is true. The literal truth of the statement is about NPs at the same level of abstraction.

A whale is a mammal.   (4.15)
A whale is like a mammal.   (4.16)

Even if the sentence is provided as a voice accompanying a picture of two whales, such that the child anchors the subject NP to one of the two whales arbitrarily, (4.15)

---

[4] Formal details of this system are available in an earlier version of this paper [29].

remains a literally true statement. As an accepted piece of news, the child extends whatever meaning of "mammal" was in place before, with the new information that one or more whales is also in that set. If the child knows that whales are not fish, the child may retract the prior creative hypothesis that the swimming fish-like thing is not a fish. Note that (4.16) is also true because whales are mammals, and things are generally like themselves.[5] Moreover, (4.16) is reversible: a mammal is like a whale for, among others, all the reasons that make the kind, whale, a sub-kind of mammals. This is the same as squares being like rectangles and rectangles being like squares. Of course, the simile isn't particularly felicitous given the truth of the stronger class inclusion statement of (4.15). Glucksberg notes that metaphors are not only asymmetric, they are also sometimes only reversible with a change of meaning into a different metaphor. [10, p. 45] notes the difference between (4.17) and (4.18).

Some surgeons are butchers.   (4.17)
Some butchers are surgeons.   (4.18)

The former presumably has negative connotations, and the latter, positive. Later the issue of reversibility returns with emphasis on the fact that the constraint is not simply on the linear presentation of topic and vehicle (see (4.34)).

Reversing (4.15), (4.19) is also felicitous – if it expresses that a specific kind of mammal is the kind "whale"; or if it means that a particular individual mammal is of the whale sort; or (least likely) if a specific indefinite is both a mammal and a whale.

A mammal is a whale.   (4.19)

These properties of generics indicate that plurality of reference, the possibility of words being used in strict or loose senses with graduated literalness, with access to individuals and their kinds, is not unique to metaphorical expressions.

The point of the example (4.15) is to emphasize that there are needs for asserting and retracting information about entities and relationships that hold among entities in the world, independently of whether the utterance accepted as effecting the change fits criteria for some figure of speech or other. A mechanism for assertion and retraction is a necessary part of information processing.

4.2.2 A Formal Model of First-Order Belief Revision

Oliver Lemon [22] provided a framework for modeling first-order belief revision of incomplete theories. A theory in this framework is a set of agent beliefs about the world and the individuals and first-order relations within it. An agent can obtain new beliefs or retract old ones. Beliefs may be about the truth of propositions or of properties holding of named individuals. A common simplifying assumption is made that every individual in the domain has a name [8]. Additional beliefs may include

---

[5] It is felicitous for someone to say, "He is not like himself today."

quantificational statements, and in fact may be about any well formed sentence in a standard first order language. Beliefs, quantificational or not, may be added or subtracted. Rationality postulates ensure consistent belief states under deductive closure.

In retracting a belief from a theory, in general there will not be a unique sub-theory of T that fails to entail the retracted formula (e.g. ϕ). Lemon refers to max-imal sub-theories of T with that status as, $T \perp \phi$, and defines a choice function α to pick out members of that set, and an intersection over all possible choices yields a total retraction of the formula ϕ from the theory T. To retract a universally quan-tified formula involves total retraction of a single formula in which the quantifier is removed and free instances of the erstwhile bound variable are substituted with a constant, the name of the individual which causes the universal to be retracted. Total retraction of an existentially quantified formula similarly requires retraction of all formulas obtained by substitution of each constant for now free instances of the formerly bound variable. This method adopts a substitutional approach to quantification. Names are taken as rigid designators and the naming of individuals in the domain is only ever monotonically increasing – it is not possible to un-name an individual, although individuals may have more than one name.

4.2.3 First-Order Belief Revision Adapted to Sense Extension

In general, dynamic semantics supposes that there is an input to interpretation and that the output of interpretation can be a truth value, but also a change in the model of the world that is input to interpretation of subsequent utterances. In classical logic, one thinks of a meaning function defined for arbitrary sentences relativized to a model which consists of a domain and interpretation function. In an exten-sional semantic analysis, the interpretation of a predicate is the set of tuples each of which the predicate is true of; the interpretation of a constant is some element of the domain. Updating or downdating means adding tuples to or subtracting tuples from the interpretation function. Additional parameters are needed for interpretation to accommodate multiple senses. Two additional aspects of context also anchor the interpretation – the default sense of an expression and the default "world" in which interpretation is happening.[6] Assuming a fixed domain, with dynamic interpretation, relativization is to the input and output interpretation function. Thus, a basic mean-ing function is annotated with the input and output interpretation functions (as well as assignment functions for free variables – these function like contexts that provide the reference of pronouns), accordingly. With static interpretation, the inputs and outputs are identical. For dynamic interpretation, the interpretation function may expanded and contract. The construction stipulates what arbitrary sentences of the language should mean; this is spelled out recursively with cases for each connective.

---

[6] An article in *The Economist* may use without penalty "bank" in an article reviewing property values on one side of the Seine.

4.2.3.1 Sense-Relative Interpretation

Interpretation is relative to models consisting of a domain of entities and an interpretation function for basic expressions in the language, which is presented in terms of the tuples comprising it. An important parameter of interpretation function is the index at which a basic expression is to be interpreted. The language supplements first-order systems as standardly presented by including expressions that designate the indices at which predications and constants are to be interpreted. While this not the way first-order systems are typically presented, it entails no more than first order expressivity: it is tantamount to having $bank^{GEOLOGY}$ and $bank^{FINANCE}$ as well-formed predicate names, where a predicate name is disambiguated with a designation of sense. Constants may similarly be accompanied by designations of sense (as a model of deixis accompanying natural language, for example). In both cases, a default sense may be assumed, and interpreting a sequence of expressions may involve changing the index at which constituent expressions are evaluated.

The system also includes the possibility of choosing between static and dynamic interpretation of expressions. Static interpretation involves inspecting what a constant refers to or testing the truth of a predication at an index. Dynamic interpretation involves either contraction or expansion: either a predication has its meaning reduced at an index so that it applies to fewer entities (or sequences of entities, depending on the arity of the predicate), or a predication has its meaning expanded at an index to apply to more entities.

4.2.3.2 Sense-Relative Assertion

In an initial proposal for analyzing metaphor with dynamic semantics, static interpretation was reserved for senses classified as literal and dynamic interpretation for senses classified as non-literal [27]. What is correct about this distinction is that the difference between a literal sense and a non-literal sense is convention in classifying it as such. Here, a partial ordering in that dimension is assumed (this emerges more below, particularly in how this relates to genericity). Evidently, people are able to perceive degrees of metaphoricity [24]. I argue that this approach is incorrect in providing belief revision only for non-literal expressions; the independent need for sense extension and contraction was motivated in Section 4.2.1

"Constants" can be supplied with new senses and references within those senses. The interpretation of a tuple of such terms requires passing the output of the interpretation of one argument into the input of interpretation of the following one. This idealization is too strong, in general, because it works on canonical argument structure, without taking into account non-canonical orderings of argument realization, through topicalization, for example. The assertional interpretation of a predication (or proposition) always succeeds relative to either a designated or default sense. It has the effect of adding a tuple (possibly empty for a proposition) to the characteristic function for the n-ary predicate for the relevant sense.

By construction, the assertional interpretation, if repeated for sufficient designations of elements of the domain, can come to make the static interpretation of the universal quantifier work out to be true, and it can make existential generalizations true in a single application for the relevant sense. In the case of static interpretation (Section 4.2.3.1) implication and disjunction require no mention because they are defined from negation and implication. In the case of dynamic interpretation, those connectives are constrained to be static. Conjunction is given a dynamic interpretation: giving an assertional interpretation to an initial conjunct changes yields a change in the background model that serves as interpretation input to a subsequent conjunct. Thus, conjunction is not certain to be commutative in the dynamic fragment. The dynamic fragment is non-monotonic.

Further, in the underlying formal system there is no direct clause for extending the sense of a predicate within the scope of a quantifier, but doing so with individual constant terms will have the effect of making static interpretation relative to the selected sense work out to be true. Similarly, senses of predicate names and constants cannot, by this construction, be augmented under the scope of negation. However, because extension of a predicate at an index for a sense provides grounds for static interpretation of an existential generalization to be true, it equally supplies grounds for a formerly true negated existential generalization to be false. Even just addition of truths inside the model yields non-monotonicity.

4.2.3.3 Sense-Relative Retraction

I assume that names of individuals cannot be retracted.[7] Thus, names and tuples of names will be interpreted as what they mean according to a static designated sense. The output of retracting information about a particular tuple of individuals from the denotation of a predicate for some sense of the predicate is an interpretation function which is smaller (if that tuple was in the background model for the predicate at that sense in the first place), and the formula will evaluate to be false. Subsequent static interpretation of the negated formula, picking out exactly that same tuple, will evaluate as true because the non-negated form is now false.

Universally quantified formulas (possibly complex) may be retracted by deleting a tuple from the interpretation function that creates an exception. Existentially quantified formulas may be retracted by deleting all tuples that support the existential generalization. The only generalization over Lemon's work assumed in this section is that retraction of information is relativized to the sense of the predicate at stake. It uses an extensional unpacking of intensions.

4.3 Ramifications for Metaphoricity

The discussion which precedes has not provided the logic which fits the constraints on updating and downdating models as specified. Ensuring the correspondence between alterations to models and closure of the set of sentences true in those

---

[7] This is not an assumption without precedent [22].

models is a separate exercise. However, it can be seen from what is discussed what sentences will gain or lose support and that the entire system is non-monotonic, because the underlying models are non-monotonic: relations can expand and contract. The location of dynamic semantics for the language is in the non-logical expressions – proposition and predicate names as well as names of individuals (all relative to senses of them). It is possible to imagine varying the interpretation of the logical operators ($\wedge$, $\neg$, etc.) so that they do not behave in classical ways orthogonally to dynamism [20]; however, that is not of focus here. The language is set up such that in NPs, head noun restrictor sets; in VPs, verbal heads; in APs, adjectives and adverbs; in PPs, prepositions may expand and contract the sets that they are true of as individuals or tuples of individuals corresponding to relations.

It is assumed that these sets are the input to generalized quantifier constructions [2] to, for example, construct an NP as a set of sets which "lives on" its head noun set, and such that a positive polarity sentence involving an NP and an intransitive VP or copula-linked predication is true just if the set given by the predicate is an element of the set of sets provided by the NP. If metaphorical statements are taken to be class inclusion statements, this analysis in terms of generalized quantifiers will demand modification to achieve the same effect. In fact, the inclusion statement is that the "lives on" property holds: whether the characteristic set χ corresponding to any predicate is an element of the quantifier depends only on the intersection of the head noun set (N) from the quantifier with χ. For any χ that is in the GQ denotation supersets or subsets will either have to also be elements of the GQ denotation as well (or must not be) depending on the determiner that combines with the head noun set to form the GQ. Thus, the "lives on" property takes care of class inclusion, but also exclusions where necessary. The reason to accept generalized quantifier theory is its robust account of evidently syntactic puzzles (e.g. the "definiteness effect" in partitive constructions), semantic puzzles (e.g. licensing of negative polarity items by downwards monotone determiners), as well as predicting processing facts about natural language determiners (e.g. monotonic increasing determiners (e.g. "some" and "all") are easier to evaluate than monotone decreasing determiners (e.g. "no" and "few"), which are in turn easier than non-monotonic determiners (e.g. "exactly three")) that are supported by empirical evidence [23]. Ample reason to move to a generalize quantifier account are provided by [2]; primary is that first-order logic does not have the expressive capacity to represent the meaning of "counting" as is required by relatively mundane natural language determiners like "most" or "many".[8] Finally, in presenting the invariants associated with generalized quantifiers, [2] assumed a fixed-model constraint to address the variance in determiner meaning that depends on contextual factors like expectations. For example, a different number of people, even a different proportion of a relevant head noun set being quantified over, might count as "many" depending on the expectations. The fact, that

---

[8] Note that [10, p. 22] recalls experiments from 1982 to 1989 which revealed significant differences in responses to metaphorical statements with quantified subjects depending on the determiner of quantification ("some" vs. "all"); one might anticipate that a wide range of variability is indexed by exactly the monotonicity properties of the determiner.

the cardinality or ratio involved in "many" is to be interpreted with varying models in generalized quantifier theory is a background support for the kind of variation in interpretation depending on signaled sense to account for aspects of metaphoricity in this paper. Consider the highlighted portion of (4.20).[9]

There was never a solicitation for money at these events, but of course, the President hoped that people in this category of friends and prior supporters would give money afterwards. *And, in fact, many did, and many did not.*   (4.20)

It is clear that metaphoricity is handled here by classification of senses of predicates as metaphorical or not, and degrees of metaphoricity can be represented. It remains to discuss more about the nature of the distinct senses of predicates and what makes them stand in special relationships to their base forms. The basic idea is that by addressing predicates and their related senses, one has access to a larger characteristic function for the set than is relevant to any literal sense of the predicate. Each possible sense is the characteristic function corresponding to an abstraction over salient properties associated with the characteristic function for the predicate. "Duality of reference" in Glucksberg's terms is a species of polysemy in which a predicate name can pick out its literal sense, or be used as a metaphor, picking out an otherwise un-named superordinate concept or category at a level of abstraction determined by the context of use (Chapter 1 by Sam Gluchsberg, 2011, this volume). There can be any number of such abstractions, and one does not expect each of them to have a unique name [10]. As constructed here, each additional sense of a predicate has its own characteristic function, and as has been seen, the set determined by each such function can be expanded or contracted using the dynamic interpretation mechanisms specified above. Equivalence classes of senses of a predicate form the space of polysemy for a predicate (as distinguished from its having unrelated homonymic senses), and all of the tuples in the entire equivalence class form a larger set than those in the basic literal sense.

The framework is outlined as above with extensional treatment of types. As such, the system can also be compared with the work of [21], who presents a framework in which linguistic tokens paired with situations appropriate for use (relativized to speakers) can be seen as individuating senses of the tokens that modulate through dialogue, addressing the kinds of circumstances that shape meaning changes.

4.4 Metaphoricity and Genericity

As constructed, predicates cannot be extended to cover new tuples under the scope of negation, but negations can be made true by retracting tuples from the characteristic functions of particular senses of predicate names. It is tempting to say that novel use of metaphor involves the generation and population of new senses of predicates;

---

[9] Attributed to Lanny Davis, special White House counsel, February 25, 1997. OnLine Focus interview with Elizabeth Farnsworth (http://www.pbs.org/newshour/bb/white_house/february97/davis_2-25.html) — last verified January 2011.

conventionalized metaphor is about the re-use of old senses, and dead metaphor does not even involve extending the predicate to a fresh set of tuples. However, it is key that information assertion and retraction about individuals and tuples of individuals is independent of metaphoricity. It happens with literal information also.

Discriminating sense makes it possible to consider subsets of the interpretation function as bundling predicates together by senses that are shared. For example, there is a financial institution sense of "bank" that is in common with a particular sense of "bond". The two words do not mean the same thing: even relative to that shared sense the words participate in different networks of implications and are true of different tuples. A partial order relation names paired with their senses provides a cline of metaphoricity. The different senses of predicates will generally be true of differing sets of tuples, and metaphorical denotations tend to be disjoint from literal counterparts. The total union of sense denotations for a predicate, a single "loosely speaking" version, has more than a constituent literal denotation univocally.

Genericity provides an alternative sense to predicates that has nearly identical properties to metaphorical sentences, but on the analysis provide here, they are explained by appeal to construction of related contracted senses of predicates. Like metaphors, generics can be predications over nominals (4.21), (4.22), and (4.23) or can involve the verbs directly as well (4.24). Generics certainly cannot be understood as universally quantified statements in a classical framework, as their nature is to have exceptions. Thus, if generics are taken to be category inclusion statements, they turn out to be false in their strict literal senses. However, generics cannot be truthfully understood as asserting even that *most* of the entities in the subject NPs head noun set have the predicated property, because (4.21), (4.22), and (4.23) would remain true if there tend to be more male platypuses than female ones, or even if most platypuses die before reaching the age of being able to reproduce. Similarly, (4.24) might be uttered to mean that the only time Leslie smokes, it's after dinner, or among the times that Leslie smokes, after dinner times are included. The safest "strong" reading of a generic in first-order languages is that the sentences make an existential claim that, for example, there is at least one platypus that has produced an egg. Equivalently, one can appeal to a universal claim over a set that has only one element, essentially evaluating the predicate at an index where the denotation is small enough to have no counter-examples. The simple existential readings are a challenge for sentences like (4.25) in which there is no real entity in the domain that satisfies the existential generalization, but universal quantification at an index where the domain is empty appears satisfactory.[10]

The platypus is an egg laying mammal.	(4.21)
A platypus is an egg laying mammal.	(4.22)
Platypuses are egg laying mammals.	(4.23)
Leslie smokes after dinner.	(4.24)
Unicorns are white.	(4.25)
An egg laying mammal is the platypus.	(4.26)

---

[10] However, at those indices, "unicorns are not white" is also true.

The truth conditions of generics are as troubled as those of metaphors. Reversing the predications is possible, but changes the meaning slightly, admitting a Gricean implicature in (4.26) that there are other egg laying mammals as well. This reversibility is comparable to that mentioned above for metaphors (recall (4.17) and (4.18)).

Returning to negation, (4.27) was discussed above (4.10) as containing sentential negation. It may also be understood as expressing a more local negation synonymous with (4.28). In this case it retains interpretation as a generic. In the account proposed, the evaluation of both sentences involves recourse to designated senses of the predications "not Japanese" and "sumo wrestlers" such that the instances of the latter are all among the former. Deep analysis of the predicates establishes the synonymy of "not Japanese" and "Gaijin" using meaning postulates (see e.g. (Chapter 3 by Jerry R. Hobbs and Andrew S. Gordon, 2011, this volume)).

Sumo wrestlers are not Japanese.   (4.27)
Sumo wrestlers are *Gaijin*.   (4.28)
Dodos are extinct.   (4.29)
Dodos are no longer living.   (4.30)

Similarly, purely kind-level predications such as (4.29) can can be addressed in extensional terms as in other approaches to generics [14].[11] Predications of kinds may be seen as equivalent to related predications of instances of the kinds. Kinds can be constructed from classes of available extensions of the corresponding predicates expanded at some indices of evaluation and contracted at others. The effects associated with "duality" of reference between kinds and their instances are attributed to picking some index or other for evaluation on one hand, or on the other hand, considering a collection of indices versus a particular index within the same sense: a plurality of reference is available. However, reification of kinds (or any other abstract notion) as the potential referents is not antithetical to the programmatic analysis argued here.

As mentioned above, indices for the evaluation of senses of predicates can be grouped according to semantic fields (so that, for example, $instrument^{FINANCE}$ may be preferred over $instrument^{MUSIC}$ when evaluating a sentence that has a prior mention of $bank^{FINANCE}$). This incorporates insights from the field of cognitive linguistics in which conceptual metaphors deliver families of predicates interpreted according to the same designated indices for evaluation (Chapter 2 by Andrew Goatly, 2011, this volume). Simultaneous ordering of indices according to categories orthogonal to semantic field, such as partial orderings by degrees and kinds of affect are also possible – this is in the spirit of the analysis of modality provided by [18] with a double-ordering of "possible worlds" according a modal base (that determines which sort of modality) and an ordering source which provides

---

[11] That work is mainly concerned with an analysis of bare plurals as not specifying explicitly their intended quantificational force over a domain named by a predicate; by comparison, the present work can be seen as advocating universal quantification for bare plurals always, but with variation in the size of reference set depending on the sense selected for the head noun predication.

accessibility relations. Rather than encoding affect associated with predications as other object-level predications, such as proposed elsewhere (Chapter 3 by Jerry R. Hobbs and Andrew S. Gordon, 2011, this volume), affect here is encoded in meta-level classification of senses.

It is common to understand generics as involving a restricted domain of quantification over salient individuals. This is the converse of what happens with metaphor understanding. Thus, the proposal to unify the treatment of metaphoricity and genericity in this dynamic framework is to allow for alternative senses of literal predicates which are reduced by individuals or tuples[12] that challenge the literal truth of universal quantification over the full domain. Metaphors are class inclusion statements that involve expanding hitherto un-named categories, and generics are class inclusion statements that involve shrinking categories with prior names. Among the alternative senses for predicates are those which stand systematically in this way via relevant restriction over the characteristic set of the predicate at some sense.

4.5 Particulars of the Class-Inclusion Framework

One aspect of the system that merits discussion is its relationship to the theory developed by Glucksberg and his colleagues. There is some divergence with respect to the question of asymmetry of metaphor, which I argued above extends somewhat to genericity. The divergence is in that the system doesn't place great emphasis on the asymmetry beyond the order of arguments in a tuple, which is in each case an ordered sequence. The system, through multiplicity of senses for predicates and terms, admits duality of reference, but it is not prejudiced to require that the dual argument must be in a non-subject position. Interestingly, [10] comments in a number of places less on the asymmetry of subject and object, as with respect to new and given. This is also called the topic-comment distinction, and it often in English coincides with the grammatical subject, but it is not analytically identical [20].

Einstein [my brother points at a clever companion] can work out how the remote control works. (4.31)

It is sharks that lawyers are. (4.32)

Sharks, Lawyers are. (4.33)

First of all, (4.31) shows that the Demjanjuk examples of [10, p. 40] involving abstract categories can occur in subject position. The cleft (4.32) and topicalization (4.33) are both constructions that move canonical objects into a topic position for information packaging purposes, and in these cases it turns out to be the abstract category that form topic, and the finite sentence with an object gap that forms a predication for the comment. Perhaps one would want to argue that the subject remains given in these and related constructions, but it is clear that it is not the linear order of presentation that matters as much as the information packaging into topic and comment.

---

[12] Individuals are singleton tuples, anyway.

However, a more robust class of examples of non-literal expressions best understood as class inclusion statements, but with the class in the initial position, has an exemplar in (4.34).[13] This construction relates directly to predication metaphor (4.35). A counterpart construction for simile is perhaps anomalous (4.37).

"Anyone who has lived in the ethnic shouting match that is New York City knows exactly what I mean"   (4.34)
New York City is an ethnic shouting match.   (4.35)
Anyone who has lived in the New York City that is an ethnic shouting match knows exactly what I mean.   (4.36)
the jail that is like Sandy's job   (4.37)

In (4.34) both terms of the predication can be understood via literal referent or as concepts, but there is evidently a preference for "the ethnic shouting match" to be understood as a name for category which is asserted to have the literal New York City within it. The relevant non-literal constituent of (4.34) can be equally understood via (4.35). An adapted formulation is provided in (4.36) to show that reversibility does obtain and "New York City" does not appear to be forced into a sub-kind level expression, although it has to be at least a category here for the definite reference to work. The point is that there is more to explore about the asymmetry facts associated with metaphor. They appear to be not simply about the order of presentation of topic and vehicle and their reversibility. The facts seem to depend upon the construction which is used to package the relevant information.

In the underlying formal system here, a sequence of arguments to a predicate is assumed to be interpreted in the order given. Where interpretation is dynamic, the interpretation function that results as the output of processing the first argument is the input to the second, and so on. The tuples are ordered by the argument structure of the predicate, rather than the information packaging of the construction it appears in. There may well be empirical consequences that depend on alternative information packaging associated with argument terms, but it is not clear that they have much significance. That is, while a tendency to restrict reversibility of arguments and correlation with topic-comment structures may be useful diagnostics of metaphoricity, the dual reference theory seems to be able to stand up independently in cases where the data seems slightly at odds with the asymmetry claims.

In particular, the dual reference theory provides an intuitive explanation for the fact that similes can be restated in stronger term as metaphors, and for the (non-universal) potential for metaphor to be paraphrased with simile, evidently shifting between non-literal and literal senses of a predicates. (Chapter 1 by Sam Glucksberg, 2011, this volume).

Sumo wrestlers are like elephants.   (4.38)
Sumo wrestlers are like Japanese people.   (4.39)
Squares are four sided equiangular polygons.   (4.40)
Squares are like four sided equiangular polygons.   (4.41)

---

[13] Attributed to Andrew Sullivan by [25].

On Glucksberg's theory, sentences involving nominal metaphors are class inclusion statements that refer to a category superordinate to the literal sense of a named predicate, where the superordinate level is determined as appropriate in the context of use. A simile can be re-expressed as a metaphor that makes use of the superordinate category as the sense of the predication. Only those sentences which are interpreted using the sense provided by the dual reference can be felicitously paraphrased with a simile. Contrast the capacity of (4.1) from the outset of the paper to be paraphrased with (4.38) and the inappropriateness of (4.41) as a paraphrase for (4.40): literal class inclusion statements do not involve dual reference in Glucksberg's sense. However, consider (4.3) in relation to (4.39) – even changing the predicate adjective in the generic to a predicative nominal for parallelism in (4.39) doesn't improve the formulation with explicit comparison. Generics function as class inclusion statements also. However, above in Section 4.4 it was argued that generics make dual reference to categories as well, but, to subordinate categories. This clarifies part of the force of Glucksberg's sense of dual reference: it is not just polysemy between a category and a hierarchically related one; rather, it crucially involves a category superordinate to the literal sense. Subordinate categories lacking prior names, universal quantification over which supports the truth of their generics, in contrast to the superordinate categories in the case of metaphors, do not participate in the all the same effects. Whereas the superordinate categories can lead to more emergent associations in the comparative constructions constituted by similes, the subordinate categories of generics necessarily yield tautologies in combination with the predication. These resist emergent associations, and are thus extremely odd.

It is important that the formal framework outlined in Section 4.2.3 addresses more than nominal metaphors linked by copular verbs which comprise the primary focus of (Chapter 1 by Sam Glucksberg, 2011, this volume). An example like (4.42) does not evidently make recourse to superordinate categories for either the subject or object nominal, nor does it obviously convey a class-inclusion statement, but it does involve dual reference with a (metaphorical) superordinate sense of "eats". As before, static interpretation can be used to evaluate the the statement as a contingent declarative, or dynamic interpretation can be used to assert its truth, updating the interpretation function. Of course, the formal details require elaboration to capture even an extensional interpretation of the verbal noun subject and the mass noun object in this example. A richer type-theoretic system such as that described by [6] will ultimately be necessary.

Covering news in the field eats money.[14]   (4.42)

Sal smokes a Cuban cigar.   (4.43)

Similarly, the habitual in (4.43) is interpreted via selection of a sense of "smokes" that refers to a category subordinate to the literal sense in terms of the quantification involving Sal and cigars – it has a smaller extension where universal quantification holds. Moreover, there are a number of such subordinate senses corresponding to

---

[14] Attributed to George F. Will by the American Heritage Dictionary.

the ways in which the habitual is to be understood (e.g. Sal prefers Cuban cigars; when Sal smokes a cigar, it is a Cuban one; there is only one type of Cuban cigar that Sal smokes; etc.). Interpretation of both is mediated by hierarchically related senses. Perhaps because neither is constructed as a class inclusion statement that could yield a tautology, both the metaphor and the habitual support reformulation as explicit comparison statements as in (4.44) and (4.45).

Covering news in the field consumes money like termites eat wood.   (4.44)

Sal consumes a Cuban cigar like Bond drinks a shaken martini.   (4.45)

In any case, the dual reference constraint between nominal metaphors in class inclusion statements and paraphrase with similes is not available for metaphorical verbs.

4.6 Final Remarks

This paper has argued that metaphoricity and genericity are best handled within the same semantic framework, one that admits information update, names of individuals and predications paired with senses. The formal machinery has been sketched in an extensional unpacking of the main ideas. Pairs of predicate names and senses can be partially ordered to achieve a continuum of metaphoricity. They may also be classified according to other meta-linguistic categories, affect among them. Sam Glucksberg (Chapter 1 by Sam Glucksberg, 2011, this volume) has argued that metaphors are best analyzed as class inclusion statements involving dual reference. Generics and habituals certainly look like class inclusion statements and show many of the same properties of non-literal interpretation that metaphors do. It has been shown exactly how metaphors relate to each other within a non-monotonic system for information change.

Acknowledgments This research is supported by Science Foundation Ireland RFP 05/RF/CMS002. I am grateful for constructive feedback from Khurshid Ahmad and the anonymous reviewers.

References

1. Alchourrón, C.E., P. Gärdenfors, and D. Makinson. 1985. On the logic of theory change: Partial meet contraction and revision functions. *The Journal of Symbolic Logic* 50:510–530.
2. Barwise, Jon, and Robin Cooper. 1981. Generalized quantifiers and natural language. *Linguistics and Philosophy* 4(2):159–219.
3. Carlson, Greg, and Jeff Pelletier, eds. 1995. *The generic book*. Chicago, IL: University of Chicago Press.
4. Cohen, Ariel. 1999. *Think generic! The meaning and use of generic sentences*. Stanford, CA: CSLI Publications.
5. Cohen, Ariel. 2001. On the generic use of indefinite singulars. *Journal of Semantics* 18(3):183–209.
6. Cooper, Robin. 2007. Copredication, dynamic generalized quantification and lexical innovation by coercion. In *4th international workshop on Generative Approaches to the Lexicon*, eds. Pierrette Bouillon, Laurence Danlos, and Kyoko Kanzaki. Available at http://www.ling.gu.se/~cooper/records/copredinnov.pdf.

7. Davidson, Donald. 1984. What metaphors mean. In *Inquiries into truth and interpretation*, ed. Donald Davidson, 245–264. Oxford: Oxford University Press.
8. Gamut, L.T.F. 1991. *Language, logic and meaning, Part 1: Introduction to logic*. Chicago: Chicago University Press.
9. Glasbey, Sheila. 2007. Aspectual composition in idioms. In *Recent advances in the syntax and semantics of tense, aspect and modality*, eds. Louise de Saussure, Jacques Moeschler, and Genoveva Puskas, 1–15. Berlin: Mouton de Gruyter.
10. Glucksberg, Sam. 2001. *Understanding figurative language: From metphors to idioms*. New York: Oxford University Press (With a contribution from Matthew S. McGlone).
11. Glucksberg, Sam, and Boaz Keysar. 1993. How metaphors work. In *Metaphor and thought*, 2nd ed., ed. Andrew Ortony, 357–400. Cambridge: Cambridge University Press (First Published, 1979).
12. Goatly, Andrew. 1997. *The language of metaphors*. London: Routledge.
13. Groenendijk, J., and M. Stokhof. 1991. Dynamic predicate logic. *Linguistics and Philosophy* 14:39–100.
14. Herbelot, Aurelie, and Anne Copestake. 2011. Formalising and specifying underquantification. In The 9th international workshop on Computational Semantics, Oxford.
15. Heyer, Gerhard. 1990. Semantics and knowledge represeantation in the analysis of generic descriptions. *Journal of Semantics* 7:93–110.
16. Kamp, Hans, and Uwe Reyle. 1993. *From discourse to logic*. Dordrecht: Kluwer.
17. Keenan, Edward. 1975. Towards a universal definition of 'subject'. In *Subject and topic*, ed. Charles Li, 304–333. London: Academic. Symposium on Subject and Topic, University of California, Santa Barbara.
18. Kratzer, Angelieka. 2002. The notional category of modality. In *Formal semantics: The essential readings*, eds. Paul Portner and Barbara Partee, 289–323. Oxford: Blackwell.
19. Krifka, Manfred, Francis Jeffry Pelletier, Gregory N. Carlson, Alice ter Meulen, Godehard Link, and Gennaro Chierchia. 1995. Genericity: An introduction. In *The generic book*, eds. Gregory N. Carlson and Francis Jeffry Pelletier. Chicago, IL: University of Chicago Press.
20. Kuhn, Stephen. 1981. Operator logic. *The Journal of Philosophy* 78(9):487–499.
21. Larsson, Staffan. 2008. Formalizing the dynamics of semantic systems in dialogue. In *Language in flux – dialogue coordination, language variation, change and evolution*, eds. Robin Cooper and Ruth Kempson, 1–21. Communication, Mind & Language, Volume 1. London: College Publications.
22. Lemon, Oliver. 1998. First-order theory change systems and their dynamic semantics. In *The Tbilisi symposium on logic, language and computation: Selected papers*, eds. J. Ginzburg, Z. Khasidashvili, C. Vogel, J.-J. Levy, and E. Vallduvi, 85–99. Stanford: SiLLI/CSLI Publications.
23. Moxey, Linda M., and Anthony J. Sanford. 1993. *Communicating quantities: A psychological perspective*. Hove: Lawrence Erlbaum.
24. Ortony, Andrew. 1979. Beyond literal similarity. *Psychological Review* 86(3):161–180.
25. Roberts, Sam. 2007. *Podcast: The New York brand of bias*. The New York Times. http://cityroom.blogs.nytimes.com/2007/07/26/podcast-the-new-york-brand-of-bias/ – last verified January 2011.
26. Verkuyl, Henk J. 1993. *A theory of aspectuality. Cambridge studies in linguistics*. Cambridge: Cambridge University Press.
27. Vogel, Carl. 2001. Dynamic semantics for metaphor. *Metaphor and Symbol* 16(1&2):59–74.
28. Vogel, Carl. 2008. Metaphor is generic. In *Sentiment analysis: Emotion, metaphor, ontology & terminology*, ed. Khurshid Ahmad, 21–28. Pecs: Folli.
29. Vogel, Carl, and Michelle McGillion. 2002. Genericity is conceptual, not semantic. In *Proceedings of the seventh symposium on logic and language*, eds. Gábor Alberti, Kata Balogh, and Paul Dekker, 163–172, Pécs, Hungary, 26–29 August.

Chapter 5
Affect Transfer by Metaphor for an Intelligent Conversational Agent

Alan Wallington, Rodrigo Agerri, John Barnden, Mark Lee, and Tim Rumbell

5.1 Introduction

In this paper we discuss aspects of the extraction and processing of affective information such as emotions/moods (e.g. embarrassment, hostility) and evaluations (of goodness, importance, etc.) when conveyed by metaphor in free-form textual utterances. This work follows from our experience in building upon an *edrama* system produced by an industrial partner of ours. Using this system, human users – school children, in the testing and development stage of our work – improvised around certain themes by typing in utterances for on-screen characters they play to utter (via speech bubbles).

Now, drama by its very nature involves emotional experience, and this is particularly the case in our use of the edrama system because of the nature of the themes or scenarios we used, namely 'school bullying' and a scenario involving a sufferer of a particularly embarrassing disease – Crohn's disease – discussing with friends and family whether or not to undergo an operation. The need for the extraction and processing of this emotion and affect arises because we added to the edrama the option of having a bit-part character controlled by an Intelligent Conversational Agent (ICA). This ICA can intervene and make largely contentless, interjections and responses in order to keep the conversation flowing. And, the interjections are emotionally appropriate because the underlying system can extract affect from the human controlled characters' utterances.[1,2] Whilst other ICA research has concerned itself with the conveyance of affect [15], it appears that the conveyance of affect via metaphor has been largely ignored. Indeed, relatively little work has been done on any detailed computational processing of metaphor. Major exceptions include [5, 8, 12–14].

---

[1] User-testing [25] shows that users have enjoyed using the system.
[2] The same algorithms are also used for influencing the characters' gesturing when a 3D animation mode produced by one of our industrial partners is used.

J. Barnden (✉)
School of Computer Science, University of Birmingham, Birmingham B152TT, UK
e-mail: J.A.Barnden@cs.bham.ac.uk

K. Ahmad (ed.), *Affective Computing and Sentiment Analysis*, Text, Speech
and Language Technology 45, DOI 10.1007/978-94-007-1757-2_5,
© Springer Science+Business Media B.V. 2011

The background to the work on the conveyance of affect via metaphor comes from the authors' approach to, and partially implemented system (ATT-Meta) for, the processing and understanding of metaphor in general [1, 2, 22]. This is a more ambitious aim than the mere recognition of a metaphor or the classification of a metaphor into one of a number of different metaphor classes or conceptual metaphors (see [13]). The details of the implemented system need not concern us since they are not used in the control of the edrama ICA. However, particular important aspects of the ATT-Meta approach, if not the ATT-Meta system, are used in this much more applied, edrama, system. Thus, our metaphor approach and system emphasizes the open-endedness of metaphorical expressions, whereby conventional metaphors and fixed phraseology may be varied, extended and elaborated upon so as to convey further information and connotations not conveyed by the conventional metaphor. Although our ICA work uses WordNet for analysis of many of the affect-conveying metaphorical senses we find, we can analyse some phrasal variation in the words and deal with some senses that are not found.

Along with, and relating to, our stress on the potential open-endedness of many metaphorical expressions, our approach and system eschews large sets of correspondences between ontologically complex source and target domains in the manner of Lakoff and Johnson's [11] 'Conceptual Metaphor Theory' e.g. ARGUMENT IS WAR, or ANGER IS HOT LIQUID UNDER PRESSURE (see [7, 10]), with the meaning of a metaphorical utterance 'read off' from the source-target correspondences. Instead we assume very few, more abstract, source-target links between specific domains and account for much of the apparent systematic relatedness between source and target domains by noting that certain types of information, relations, attributes that can be inferred as holding of the situation apparently being described under the source or literal interpretation of a metaphorical utterance transfer in an essentially invariant manner from the source, where they were inferred, to the target. This is done via a limited number of what we term View-Neutral Mapping Adjuncts or VNMAs, with a particular VNMA for each type of information that transfers in this invariant manner. For example, we assume that if a causal link can be inferred as holding between entities in the source, then the causal link will hold by default in the target, transferred to the target via the CAUSATION VNMA. Similarly, if something can be viewed as applying to a particular degree in the source, e.g. slowly, then by virtue of the DEGREE VNMA, its target equivalent will be understood as applying to the same degree and likewise with such information as duration, temporal ordering, logical relations between entities, and others. Crucially for our edrama ICA, we note two particular VNMAs. Any emotional state that is invoked either by some aspect of the source, or that holds within the source, will carry over to the target by virtue of the EMOTIONAL STATE VNMA. We also assume that a value judgement concerning something in the source will also carry over by default to the target via the VALUE JUDGEMENT VNMA. For example consider a situation in which it is said of some foul mouthed character, 'Tom is a sewer'. This can be partially analysed in terms of Reddy's [16] well known 'conduit metaphor', in which information and utterances are viewed as if passing along a conduit from speaker to hearer, but what about the negative value judgement denoted by 'sewer'?

Crucially no source-target correspondence specific to the conduit metaphor will be required for this information. Instead, the negative value judgement about the nature of the material passing through a sewer should be transferred by the Value Judgement VNMA to become a negative value judgement about the words being used or spoken. A similar negative value judgement is conveyed by 'smelly attitude' or by the comment 'you buy your clothes at the rag market', two examples taken from transcripts the system automatically recorded during user-testing.

Now, within metaphor theory, it has long been noted that emotional states and behaviour are often described metaphorically [6, 10], as in 'He was boiling inside' [feelings of anger] or 'He was hungry for her' [feelings of lust] and conceptual metaphors such as the above mentioned ANGER IS HOT LIQUID UNDER PRESSURE or LUST IS HUNGER proposed to account for this. However, the conveyance of emotion via VNMAs is of a different type to such specific metaphors, and in an analysis of the transcripts from our user-testing, the type of affect laden metaphor that motivates the emotional state and value-judgement VNMAs was found to be a significant issue in edrama: at a conservative estimate, at least one in every 16 speech-turns has contained such a metaphor (each turn is 100 characters, and rarely more than one sentence; 33,000 words across all transcripts).

We shall now discuss how our system implements the transfer of affect in a very limited range of metaphors. However, it should be noted that the system underlying our edrama ICA does not detect affect solely or even primarily via metaphor. Quite apart from the recognition of specifically emotive and affective lexis, the system deals with letter and punctuation repetition for emphasis ('yeessss,' '!!!!'), interjections and onomatopoeia (textitgrrrrrrr) (see [25] for details). Although, note that these too may be viewed as manifestations of an abstract conceptual metaphor that views or conceptualises 'more of some thing or some quality' as 'an increase along one salient dimension'; typically height. This often gives us the Lakovian conceptual metaphor MORE IS UP, but gives word length when dealing with text. The degree of increase is conveyed by our DEGREE VNMA.

Finally, note that our system uses a blackboard architecture, in which hypotheses arising from the processing go onto a central blackboard. The production of the various hypotheses can then be influenced by hypotheses posted by other processes, etc. In particular, we envisage metaphor processing being refined by using such information (see [18] for more details).

5.2 Affect via Metaphor in an ICA

As noted, our system currently only applies to a limited type of affect transfer; it detects and analyses the transference of affect in cases where a human is metaphorically cast as a non-human of various sorts, as in the following cases:

1. Casting someone as an animal. This often transfers some affect – negative or positive – from the animal to the human. Interestingly, since our attitude towards young or baby forms, regardless of the animal concerned, are typically

affectionate, affection is often transferred, even when the adult form is negative ('pig: piglet', 'dog: puppy' etc.) and this seems quite productive. We deal with animal words that have a conventional metaphorical sense but also with those that do not, for it may still be possible to note a particular affective connotation, and even if not, one can plausibly infer that some affect or other is being expressed without knowing if positive or negative.

2. Relatedly, casting someone as a monster, mythical creature or supernatural being of some sort, using words such as 'monster' itself, 'dragon,' 'angel,' 'devil.'
3. Relatedly, casting someone as an artefact, substance or natural object, as in 'Tom is a [sewer; real diamond; rock].'

We currently do not deal with the related case of casting someone metaphorically as a special type of human, using words such as 'baby,' 'freak,' 'girl' [to a boy, or to a woman], 'lunatic'.

We noted the modification of the affect conveyed by animal metaphors when the young form is used. In addition, size adjectives [17] often convey affect. Thus, 'a little X' can convey affective qualities of X such as an affectionate attitude towards X, even if the X is usually negative as in 'little devils' to describe mischievous children (compare with the baby forms above), but may sometimes convey unimportance and contemptibility as in 'you little rat'. Similarly, 'big X' can convey the importance of X ('big event') or intensity of X-ness ('big bully') – and X can itself be metaphorical as in 'big baby' when said of an adult.

5.3 Metaphor Processing

The approach is split into two parts: recognition of potential metaphors; analysis of recognised elements to determine affect. Note that in some cases, e.g. using 'pig' as a negative term for a person, the metaphor analysis requires only lexical look-up (e.g., in WordNet [24]). But, not all animal words have a person sense and as noted above baby forms often change the affect as do size adjectives. Such cases motivate further processing.

5.3.1 The Recognition Component

The basis here is a list of words/phrases (see www.cs.bham.ac.uk/jab/ATT-Meta/metaphoricity-signals.html, [23]) we term 'metaphoricity signals', that often have metaphors as collocates. They include specific syntactic structures as well as lexical strings. Of course, metaphor is often not signalled and can occur in any syntactic form and not just the forms in these lists. Here, we focus on three syntactic structures, 'X is/are a Y', 'You Y' and 'like [a] Y' and on the lexical strings, 'a bit of a', 'such a' and 'look[s] like'. Note that a distinction is often made between similes and metaphors, making the third structure a simile. Our view is that (many) similes represent just a particular way of expressing an underlying metaphorical connection between X and Y and so shouldn't be treated differently from the other realisations. In the user-testing transcripts, we judged signals as actually involving metaphor in the following proportions of cases: *X is/are a Y* – 38% (18 out of 47); *you Y* – 61% (22 out of 36); *a bit of a/such a* – 40% (but tiny sample: 2 out of 5). Also: *looks like and like* – 81% (35 out of 43).

In order to detect signals, the Grammatical Relations (GR) output from the RASP parser [3] is used. This output shows typed word-pair dependencies between the words in the utterance. Using the grammatical relations from RASP has the advantage that whether or not the noun in the Y position has adjectival modifiers the GR between the verb and Y is the same. Consequently, the detection tolerates a large amount of variation, an important desiderata for metaphor. Any such modifiers are found in modifying relations and can be extracted for later analysis.

For example, the following three GRs are output for a sentence such as 'You are a pig', so allowing an 'X is a Y' signal to be detected.

```
|ncsubj|  |be+_vbr|  |you_ppy|  |_|
```

(i.e. the subject of 'are' is 'you')

```
|xcomp|  |be+_vbr|  |pig_nn1|
```

(i.e. the complement of 'are' is 'pig')

```
|det|  |pig_nn1|  |a_at1|
```

(i.e. the determiner of 'pig' is 'a')

Note that the tags 'vbr' and 'ppy' are specific to 'are' and 'you', so we also detect tags for: 'is'; for 'he', 'she' and 'it'; and for proper and common nouns, as well.

For additional confidence with this syntactic, 'X-is-a-Y', metaphoricity signal, we also detect the lexical strings 'a bit of a' and 'such a', as in 'he is such an idiot'. Note 'idiot' would have been detected as a 'Y' type metaphor, independently of 'such a', by the syntactic structure detection process.

'Such a' is found using GRs of the following type:

```
|det|  |idiot_nn1|  |an_at1|
```

(i.e. the determiner of 'idiot' is 'an'.)

```
|det|  |idiot_nn1|  |such_da|
```

(i.e. the determiner of 'idiot' is 'such')

The 'a bit of a' strings are found similarly, but cause the complication that the word 'bit' is tagged as a noun, so will be pulled out as a metaphor word by the syntactic detection processes, instead of the intended Y word. If the 'a bit of a' string is then found, we pull out the noun relating to the 'of' that relates to 'bit', in this type of GR output:

|iobj| |bit_nn1| |of_io|

|dobj| |of_io| |idiot_nn1|

The output for the 'You Y' structure is typically as in the following example:

|ncmod| |you_ppy| |idiot_nn1|

(with Y = 'idiot') making it possible to find the structure from that one relation. However, we found a problem with using RASP to detect these 'You Y' structures. It was frequently the case that RASP's 'Part of Speech' (POS) tagger will prefer to tag 'Y' as a verb if it can. For example, the word 'cow' in place of 'idiot' is tagged as a verb. In such a case, our system looks the word up in the list of tagged words that forms part of the RASP tagger. If the verb can be tagged as a noun, the tag is changed, and the metaphoricity signal is detected. Once a syntactic structure resulting from metaphoricity signals is detected, the word(s) in Y position are pulled out to be analysed.

A third metaphoricity signalling syntactic structure we use is, 'like Y'. This is found using GR's of the following type:

|dobj| |like_ii| |pig_nn1|

'like Y' is always found in this form, with the noun in question in the dobj (direct object) relation to 'like', and with an nn1 tag. This is inserted into the list of present metaphoricity signals, and an additional flag is raised if it is found in an 'X looks like Y' structure. The 'looks like' structure can be uncovered by spotting this GR:

|iobj| |look_vv0| |like_ii|

Detection of the 'looks like' structure is similar to 'such a' in that it is in addition to the main metaphoricity signal detection, in this case not only adding confidence, but also potentially altering the meaning and analysis of the metaphor.

The result of the recognition element is threefold: (1) a list of signals; (2) the X and Y nouns from the syntactic signals; (3) a list of words modifying that noun.

5.3.2 The Analysis Component

The analysis element of the processing that we shall discuss here takes the X noun (if any) and Y noun and uses WordNet 2.0 [24] to analyse them. First, we try to determine whether X refers to a person (the only case the system currently deals with). One possibility is if the X is a pronoun such as 'you'. However, we also use a specified list of proper names of characters in the drama. If X is a person, then the Y and remaining elements are analysed using WordNet's taxonomy.

The system first checks whether the Y noun in one of its synsets (or senses) is a hyponym of (or member of the class of) animals, supernatural beings, substances, artefacts or natural objects. If this is established, then the following procedure applies.

(1) The system tries to determine whether another of the senses of the word is a hyponym of the person synset, as many such 'person as animal etc.' metaphors are already given as separate senses in WordNet. If the given word contains different synsets or senses that are hyponyms of both animal etc. and person, we search for evaluative content about the metaphor and assign the result to Y.

(2) If there are no additional person synsets, or if no positive or negative evaluation can be determined, then the system will look down the various hyponym chains of the word to find all instances that also have a sense that can be determined to be a type of person (i.e. are hyponyms of the person synset). The evaluative content of these is then sought and a count made of the negative versus positive evaluations. Currently, if the ratio of one to the other is 3 or greater, then the Y is given a positive or negative evaluation accordingly.

(3) If this search too results in failure, then the hypernym chain of the word is searched up to the 'animal, artefact, etc, level, regardless of whether they have a person sense, and these are evaluated as either positive or negative.

So how do we determine positive or negative evaluations for the different senses? We have developed a method of automatically detecting the evaluation of a given metaphorical sense of a word. Intermediate synsets between the metaphorical sense of the given word and the person synsets (or the animal, artefact, etc synsets if all else fails) contain glosses, which are free-text descriptions of the semantic content of a synset. Take, for example, the word 'shark'. One of its synsets is a hyponym of 'person'. The gloss of this particular synset states that it is 'a person who is ruthless and greedy and dishonest'. Similarly, one synset of 'fox' has a gloss as 'a shifty deceptive person'. We search the words and glosses from the intermediate synsets for words, such as 'ruthless' or 'shifty' that indicate a particular affective evaluation. (see [21] for related use of WordNet glosses).

But, how is it known that 'ruthless', for example, has a negative evaluation? Now there exist numerous lists and resources containing evaluative words. Indeed, SentiWordNet [4] is based on the glosses of the WordNet synsets and assigns three numerical scores describing how objective, positive, and negative the terms contained in the synset are. See also WordNet-Affect [20]. However, in practice we found that very many of the animals etc. we wished to assign a positive or negative evaluation to were given a neutral score in SentiWordNet and so we created our own list. We decided that since we were searching though WordNet glosses, it would be most appropriate to create a list from WordNet itself. This we did in the following manner. WordNet contains a 'quality' synset which has 'attribute' links to four other synsets, 'good', 'bad', 'positive' and 'negative'. We are currently only looking for positive or negative affective evaluations, so this group of synsets provides a core set of affect indicating words to search for in the intermediate nodes. This set is expanded by following WordNet's 'see also' links to related words, to produce lists of positivity and negativity indicators. For example, 'bad' has 'see also' links to five synsets, including 'disobedient' and 'evil'; we then look up the 'see also' links in these five synsets and include these related words in the 'bad' list, and so on. If this procedure is followed through four iterations, it produces a list of 37 'Negative' words, 18 'Positive' words, 68 'Bad' words and 77 'Good' words.

With this list, we can search through the words and glosses from the intermediate nodes between the given metaphor synset (arising from the Y component in the sentence) and 'person', tallying the positivity and negativity indicating words found.

However, if this were all that is done, then an incorrect evaluation would be assigned when the glosses include negation with 'not'. For example, the gloss for 'persona non grata' is 'a person who for some reason is not wanted or welcome' and 'wanted' and 'welcome' are on the list of positive words. What we need instead are the terms 'unwanted' or 'unwelcome'. To deal with such cases, we employ a procedure called 'NotChecker'. This searches the gloss for structures of the type 'is not X' or 'is not X or Y'. Once it has found a structure like this, it will look in WordNet for an antonym of X (and Y, if necessary), antonyms are listed in WordNet, and the X and Y are replaced by their antonyms when the search is made for positive or negative evaluations. For example, in the 'persona non grata' case, NotChecker will detect the 'not wanted or welcome' in the gloss and the WordNet entries for 'wanted' and 'welcome' are examined for antonyms. Note to guard against multiple antonyms with different senses, the antonyms are themselves checked for antonyms to see if the original word is listed and it is this antonym that is used. An antonym of 'wanted' is 'unwanted' and so 'unwanted' is used in the evaluation search.

Given the words from the glosses, including those added via NotChecker, and the lists of affective words derived by following 'see also' links from 'good' and 'bad', we can then assign the affective evaluation of the metaphor. If there are more negativity indicators than positivity indicators, it suggests that when the word is used in a metaphor it will be negative about the target. If the numbers of positivity and negativity indicators are equal, then the metaphor is labelled positive or negative, implying that it has an affective quality but we cannot establish what. This label is also used in those examples where an animal does not have a metaphorical sense in WordNet as a kind of person (for example, 'You elephant' or 'You toad'). In these cases, case (2) of the procedure described above will be tried.

It might be thought that the need for an additional person hypernym for Y is not necessary and that case (3) of the procedure on its own will suffice, i.e. a search through the glosses of just the animal etc synsets in the hypernym tree for Y would yield a relevant affective evaluation. But this appears not to be the case. The glosses tend to be technical with few if any affective connotations. For example, 'toad' surprisingly does not have an alternative person sense in WordNet. The glosses of its 'amphibian', 'vertebrate' and 'chordate' hypernyms give technical information about habitat, breeding, skeletal structure, etc. but nothing affective. Worse still, false friends can be found. Thus, the word 'important' is used in many glosses in phrases like 'important place in the food chain' and this consequently causes some strange positive evaluations (for example of 'Cyclops' or 'water fleas'). It is for this reason that this search is carried out only as a last resort.

We noted earlier that baby animal names can often be used to give a statement a more affectionate quality. Some baby animal names such as 'piglet' do not have a metaphorical sense in WordNet. In these cases, we check the word's gloss to see if it is a young animal and what kind of animal it is (The gloss for piglet, for example, is 'a young pig'). We then process the adult animal name to seek a metaphorical meaning but add the quality of affection to the result. A higher degree of confidence is attached to the quality of affection than is attached to the positive/negative result, if any, obtained from the adult name. Other baby animal names such as 'lamb' do have a metaphorical sense in WordNet independently of the adult animal, and are therefore evaluated as above. They are also tagged as potentially expressing affection, but with a lesser degree of confidence than that gained from the metaphorical processing of the word. Finally, note that the youth of an animal is not always encoded in a single word: e.g., 'cub' may be accompanied by specification of an animal type, as in 'wolf cub'. An extension to our processing would be required to handle this and also cases like 'young wolf' or 'baby wolf'.

If any adjectival modifiers of the Y noun were recognized the analyser goes on to evaluate their contribution to the metaphor's affect. If the analyser finds that 'big' is a modifying adjective of the noun it has analysed, the metaphor is marked as being more emphatic. If 'little' is found the following is done. If the metaphor has been tagged as negative and no degree of affection has been added (from a baby animal name, currently) then 'little' is taken to be expressing contempt. If the metaphor has been tagged as positive OR a degree of affection has been added then 'little' is taken to be expressing affection. These additional labels of affection and contempt are used to imply extra positivity and negativity respectively.

5.4 Examples of the Course of Processing

In this section we discuss three examples in detail and seven more with brief notes.

5.4.1 *You Piglet*

1. The metaphor detector recognises the 'You Y' signal and puts the noun 'piglet' on the blackboard.
2. The metaphor analyser reads 'piglet' from the blackboard and detects that it is a hyponym of 'animal'.
3. 'Piglet' is not encoded with a specific metaphorical meaning ('person' is not a hypernym). So the analyser retrieves the gloss from WordNet.
4. It finds 'young' in the gloss and retrieves all of the words that follow it. In this example the gloss is 'a young pig' so 'pig' is the only following word. If more

than one word had followed, then the analysis process is repeated for each of the words following 'young' until an animal word is found.

5. The words and glosses of the intermediate nodes between 'pig' and 'person' contain 0 positivity indicating words and 5 negativity indicating words, so the metaphor is labelled with negative polarity.

6. This example would result in the metaphor being labelled as an animal metaphor which is negative but affectionate with the affection label having a higher numerical confidence weighting than the negative label.

5.4.2 Lisa Is an Angel

1. The metaphor detector recognises the 'X is a Y' signal and puts the noun 'angel' on the blackboard. 'Lisa' is recognised as a person through a list of names provided with the individual scenarios in e-drama.
2. The metaphor analyser finds angel that it is a hyponym of 'supernatural being'.
3. It finds that in another of its senses the word is a hyponym of 'person'.
4. The words and glosses of the intermediate nodes between 'angel' and 'person' contain 8 positivity indicating words and 0 negativity indicating words, so the metaphor is labelled with positive polarity.
5. This example results in the metaphor being labelled as a positive supernatural being.

5.4.3 Mayid Is a Rock

1. The metaphor detector recognises the 'X is a Y' signal and puts the noun 'rock' on the blackboard. 'Mayid' is recognised as a person through a list of names provided with the individual scenarios in e-drama.
2. The metaphor analyser finds rock is a hyponym of 'natural object'.
3. It finds that in another of its senses the word is a hyponym of 'person'.
4. The words and glosses of the intermediate nodes between 'rock' and 'person' contain 4 positivity indicating words and 1 negativity indicating words, so the metaphor is labelled with positive polarity.
5. This example would result in the metaphor being labelled as a positive natural object.

5.4.4 Other Examples

1. 'You cow': this is processed as a negative animal metaphor. The synset of 'cow' that is a hyponym of 'person' has the gloss 'a large unpleasant woman'. Interestingly, 'large' is included in the list of positivity indicators by the current compilation method, but the negativity of the metaphor is confirmed by analysis

of the intermediate synsets between 'cow' and 'person', which are 'unpleasant woman', 'unpleasant person' and 'unwelcome person'. These synsets, along with their glosses, contain six negativity and just one positivity indicator.

2. 'You little rat': this animal metaphor is determined as negative, having three senses that are hyponyms of 'person', containing three positivity indicators and five negativity indicators. 'Little' provides an added degree of contempt.

3. 'You little piggy': 'piggy' is recognized as a baby animal term and labelled as expressing affection. The evaluation of 'pig' adds a negative label, with no positivity indicators and three negativity indicators, and 'little' adds further affection since the metaphor already has this label from the baby animal recognition. This is therefore recognized as a negative metaphor but meant affectionately.

4. 'You're a lamb': recognized as an animal metaphor and a young animal. It has an 'affectionate' label and is recognized as a positive metaphor, with its two senses that are hyponyms of 'person' contributing two positivity indicators and one negativity indicator. The negative word in this case is 'evil', coming from the gloss of one of the intermediate synsets, 'innocent': 'a person who lacks knowledge of evil'. This example highlights a failing of the NotChecker procedure.

5. 'You are a monster': one sense of monster in WordNet is a hyponym of animal. Therefore, this is recognized as an animal metaphor, but affect evaluation reveals three negativity and three positivity indicators, so it is analysed as 'positive or negative'. These indicators are found in two opposed senses of monster: 'monster, fiend, ogre': 'a cruel wicked and inhuman person' (analysed as negative); and 'giant, monster, colossus': 'someone that is abnormally large and powerful' (analysed as positive, due to 'large' and 'powerful').

6. 'She's a total angel': a positive supernatural being metaphor, with eight positivity indicators and no negativity indicators from two senses that are hyponyms of 'person', but currently 'total' makes no contribution.

7. 'She is such a big fat cow': a negative animal metaphor made more intense by the presence of big. It has an extra level of confidence attached to its detection as two metaphoricity signals are present but currently 'fat' makes no contribution.

8. 'He's a reptile': reptile has as animal sense and no person sense, but has 4 negative hyponyms and no positive hyponyms, so it is evaluated as negative through the hyponym method (i.e. 2).

5.5 Results

It must be borne in mind when evaluating the system, exactly what the system described here is being used for: This is to detect and determine the affect if any in the utterances of human controlled avatars in an edrama setting, when a non-human quality is being attributed to a human. The system was not designed to return an accurate affectual score to any possible animal, artefact, etc that might be given to the system, or to populate WordNet or similar with such scores. Thus the fact that the system evaluates, perhaps counter-intuitively, water fleas as conveying positive, rather than neutral, affect is of little concern unless water fleas were likely to be used with the intention of conveying some affect in utterances such as 'You are a water flea'.

With this in mind we can report on some evaluations of the system. There are a total of 27,053 entries in WordNet for the five categories of animal, artefact, natural object, substance and spiritual being. To create a gold standard against which the performance of the system could be compared, one of the authors went through this list choosing items which were judged to have a strong positive or negative evaluation. This list was then circulated for comment or revision and a total of 63 positive and 141 negative entities decided upon.

We noted in Section 5.3.2 that we created a list of positive and negative evaluative terms by iteratively following WordNet 'see also' links from the terms 'good' and 'bad'. Clearly, the more iterations made the larger the list, but the more likely it is that some of the words on the list would not express a positive or negative evaluation. After some experimentation we settled on a list created by four iterations. We also noted in Section 5.3.2 that we would give an evaluation via the hyponym method if the ratio of positives to negatives or vice-versa was 3 to 1. Using these two parameters in the system, a total of 3,832 items were given a positive evaluation and 1,128, a negative evaluation. Of the 3,832 positive evaluations, 41 appeared on the gold standard list, i.e. 65.1% of the gold standard positive items were correctly tagged. Of the 1,128 items given a negative evaluation, 59 appeared on the gold standard list, i.e. 42% were correctly tagged. Of course, conversely of the 3,832 positive evaluations, only 1% (41) appeared on the gold standard positive list and of the 1,128 negative evaluations, only 5.2% were also on the gold standard negative list. But as noted, for our purposes these latter statistics are not very important.

We noted earlier that we had found SentiWordNet [4], which assigns numerical scores indicating how objective, positive or negative WordNet senses are, to be inadequate for our needs, and we can give some results to support this claim, by also evaluating it against our gold standard. Of the 63 positive senses, SentiWordNet only gave a higher positive than negative score to 14 of these, against 41 for our system. Indeed, it treated 4 of them as negative: 'cherub'; 'rock'; 'Christ'; 'sweet'. Items receiving no evaluation included such seemingly positive entities as 'angel' 'lion' and 'diamond'. Turning to negatives, both our system and SentiWordNet assigned a negative evaluation to 59 items. However, there were 8 items SentiWordNet marked as positive, including 'crap', 'Judas' and 'sheep.' Among the items receiving no positive or negative evaluation were 'toad', 'vermin' and 'punk'.

5.6 Conclusions and Further Work

The paper has discussed a relatively 'shallow' type of metaphor processing, although our use of robust parsing and complex processing of a thesaurus take it well beyond simple keyword approaches or bag-of-words approaches. Note that we do not wish simply to 'precompile' information about animal metaphor (etc.) by building a complete list of animals (etc.) in any particular version of WordNet (and also adding the effects of potential modifiers such as 'big' and 'little'), because we wish to allow the work to be extend to new versions of WordNet and to generalize as appropriate to ontologies and thesauri other than WordNet, and because we wish to allow ultimately for more complex modification of the Y nouns, in particular by going beyond the adjectives 'big' and 'little'. We recognize that the current counting of positive and negative indicators picked up from glosses is an over-simple approach, and that the nature of the indicators should ideally be examined. This is a matter of both ongoing and future research. The processing capabilities described make particular but nonetheless valuable and wide-ranging contributions to affect-detection for ICAs. Although designed for an edrama system, the tech-niques plausibly have wider applicability. The development of the processing in a real-life application is also enriching our basic research on metaphor, such as the role of VNMAs. In particular our use of hypernym trees in WordNet and potentially in other ontologies, chimes with contemporary views of metaphor theory that empha-sise category membership, such as the work of [9] or recent work in Relevance Theory and metaphor [19].[3] However, our search for specific types of information, as with our VNMA approach, that might be found in the hypernym trees is not found in such approaches.

Acknowledgments This work has been supported by EPSRC grant EP/C538943/1 and grant RES-328-25-0009 from the ESRC under the ESRC/EPSRC/DTI 'PACCIT' programme. We wish to thank our colleagues Catherine Smith and Sheila Glasbey.

References

1. Agerri, R., J. Barnden, M. Lee, and A. Wallington. 2007. Metaphor, inference and domain independent mappings. In *Proceedings of the international conference on Recent Advances in Natural Language Processing (RANLP 07)*, eds. G. Angelova, K. Bontcheva, Mitkov, R., and N. Nicolov, 17–24, Borovets, Bulgaria.
2. Barnden, J., S. Glasbey, M. Lee, and A. Wallington. 2004. Varieties and directions of interdo-main influence in metaphor. *Metaphor and Symbol* 19:1–30.
3. Briscoe, E., J. Carroll, and R. Watson. 2006. The second release of the RASP system. In *Proceedings of the COLING/ACL 2006 Interactive Presentation Sessions*, 77–80, Sydney.
4. Esuli, A., and F. Sebastiani. 2006. SentiWordNet: A publicly available lexical resource for opinion mining. In *Proceedings of the 5th conference on Language Resources and Evaluation (LREC 2006)*, 417–422, Genova, Italy.
5. Fass, D. 1997. *Processing metaphor and metonymy*. Greenwich, Connecticut: Ablex.
6. Fussell, S., and M. Moss. 1998. Figurative language. In *Emotional communication*, eds. S.R. Fussell and R.J. Kreuz, 113–142. Social and Cognitive Approaches to Interpersonal Commu-nication. Hillside, NJ: Lawrence Erlbaum.
7. Gibbs, R. 1992. Categorization and metaphor understanding. *Psychological Review* 99(3):572–577.
8. Hobbs, J. 1990. *Literature and cognition*. Center for the Study of Language and Information, Stanford University: CSLI Lecture Notes, 21.

---

[3] Although we did not pursue this in our work, exploring meronym and holonym links in WordNet might be of use in interpreting limited types of metonymy.

 9. Glucksberg, S., and B. Keysar. 1993. How metaphors work. In *Metaphor and thought* (2nd ed.), A. Ortony, 401–424. Cambridge, MA: Cambridge University Press.
10. Kövecses, Z. 2000. *Metaphor and emotion: Language, culture and body in human feeling*. Cambridge, MA: Cambridge University Press.
11. Lakoff, G., and M. Johnson. 1980. *Metaphors we live by*. Chicago, IL: University of Chicago Press.
12. Martin, J. 1990. *A computational model of metaphor interpretation*. San Diego, CA: Academic.
13. Mason, Z. 2004. CorMet: A computational, corpus-based conventional metaphor extraction system. *Computational Linguistics* 30(1):23–44.
14. Narayanan, S. 1999. Moving right along: A computational model of metaphoric reasoning about events. In *Proceedings of the National Conference on Artificial Intigence*, 121–127. AAAI Press.
15. Picard, R.W. 2000. *Affective computing*. Cambridge, MA: The MIT Press.
16. Reddy, M.J. 1979. The conduit metaphor: A case of frame conflict in our language about language. In *Metaphor and thought* (2nd ed. 1993), ed. A. Ortony, 164–201. Cambridge: Cambridge University Press.
17. Sharoff, S. 2006. How to handle lexical semantics in SFL: A Corpus study of purposes for using size adjectives. In *Systemic linguistics and Corpus*, eds. S. Hunston and G. Thompson, 184–205. London: Equinox.
18. Smith, C.J., T.H. Rumbell, J.A. Barnden, M.G. Lee, S.R. Glasbey, and A.M. Wallington. 2007. Affect and metaphor in an ICA: Further developments. In *Intelligent virtual agents*, eds. C. Pelachaud, J.-C. Martin, E. André, G. Chollet, K. Karpouzis, and D. Pelé, 405–406. 7th international working conference, IVA 2007. Lecture Notes in Computer Science, vol. 4722. Heidelberg: Springer.
19. Sperber, D., and D. Wilson. 2008. A deflationary account of metaphor. In *Handbook of metaphor and thought*, ed. R. Gibbs, 84–109. Cambridge: Cambridge University Press.
20. Strapparava, C., and V. Valitutti. 2004. WordNet-Affect: An affective extension of WordNet. In *Proceedings of the 4th international conference on Language Resources and Evaluation (LREC 2004)*, 1083–1086, Lisbon, Portugal.
21. Veale, T. 2003. Systematicity and the lexicon in creative metaphor. In *Proceedings of the ACL workshop on Figurative Language and the Lexicon, the 41st annual association for Computational Linguistics Conference (ACL 2003)*, ed. A.M. Wallington, 28–35. East Stroudsburg, PA: Association for Computational Linguistics.
22. Wallington, A., J. Barnden, S. Glasbey, and M. Lee. 2006. Metaphorical reasoning with an economical set of mappings. *Delta* 22(especial):147–171.
23. Wallington, A.M., J.A. Barnden, M.A. Barnden, F.J. Ferguson, and S.R. Glasbey. 2003. *Metaphoricity signals: A Corpus-based investigation*. CSRP-03-5, School of Computer Science. Birmingham: Birmingham University.
24. WordNet. 2006. *WordNet, a lexical database for the English language*. Version 2.1 Cognitive Science Laboratory. Princeton University.
25. Zhang, L., J.A. Barnden, R.J. Hendley, and A.M. Wallington. 2006. Exploitation in affect detection in improvisational E-drama. In *Proceedings of the 6th international conference on Intelligent Virtual Agents*, eds. J. Gratch, M. Young, R. Aylett, D. Ballin, and P. Olivier, 68–79. Lecture Notes in Computer Science, 4133, Springer.

Chapter 6
Detecting Uncertainty in Spoken Dialogues: An Exploratory Research for the Automatic Detection of Speaker Uncertainty by Using Prosodic Markers

Jeroen Dral, Dirk Heylen, and Rieks op den Akker

6.1 Introduction

Each utterance we make comes with a particular degree of certainty we have about the state of affairs that is described in our utterance. We may feel reasonably confident or rather hesitant about whether there is any truth in what we are saying. We often express this degree of certainty in what we are saying through hedges ('I think'), modal verbs ('might'), adverbs ('probably'), tone of voice, intonation, hesitations. Our speech may co-occur with gestures and facial expressions that can express the same hesitant or confident state of mind. This research will focus on the prosodic features of speech and will try to develop a method to automatically classify speech as being (un)certain. The purpose of this research is to (automatically) measure one's belief (or confidence or self-conviction) in the correctness of a certain utterance. Even when the definition of uncertainty is clear, a number of questions can still be posed: how to state the degree of uncertainty? Is it certain or uncertain or are there shades of grey in between? And if so, how do we state them?

6.2 Related Work

Although uncertainty can be detected by both visual and non-visual means, this research, and the overview of the related work, will focus on the non-visual aspects of the detection of (un)certainty.

6.2.1 Defining (Un)certainty

People's ability to accurately assess and monitor their own knowledge has been called the 'feeling of knowing' (FOK) by Hart [6]. Many experiments in this area are

D. Heylen (✉)
University of Twente, Enschede, The Netherlands
e-mail: heylen@ewi.utwente.nl

K. Ahmad (ed.), *Affective Computing and Sentiment Analysis*, Text, Speech
and Language Technology 45, DOI 10.1007/978-94-007-1757-2_6,
© Springer Science+Business Media B.V. 2011

based on question-answering where respondents must answer certain (knowledge) questions and assess whether their answer is likely to be correct. A study by Smith and Clark [12] investigated FOK in a conversational setting and followed the method mentioned above. Respondents were asked to answer general knowledge questions, then estimated their FOK about these questions and finally were tested on their ability to recognize the correct answer. They found that FOK was positively correlated with recognition and with response latency when retrieval failed and negatively correlated when retrieval succeeded.

Another study by Brennan and Williams [5] used the research of Smith and Clark and in addition researched the sensitivity of listeners to the intonation of answers, latencies to responses and the form of non-answers. When looking at the 'feeling of another's knowing' or FOAK, Brennan and Williams state a listener can use several different sources of information to evaluate a respondent's knowledge:

- His/her own knowledge;
- Assess the difficulty of the question for the average person or for the typical member of a particular community and use that information to judge a respondent's confidence;
- Information from their shared physical environment and from immediately previous conversation – "mutual knowledge";
- Information about the respondent's ability or previous performance;
- Paralinguistic information displayed in the surface features of respondent's responses (intonation, latency to response).

In their experiments Brennan and Williams concentrated on the paralinguistic information available. The result of their experiments supports the interactive model of question-answering and perhaps helps in understanding respondent's metacognitive states when searching their memories for an answer. Brennan and Williams' work also demonstrates the ability of listeners to use these cues. Their FOAK was affected by the intonation of answers, the form of non-answers and the latency to response (e.g. a rising intonation often accompanied a wrong answer).

Krahmer and Swerts [7] describe experiments with adults and children on signaling and detecting of uncertainty in audiovisual speech. They found that when adults feel uncertain about their answer they are more likely to produce pauses, delays and higher intonation (as well as some visual signals, such as eyebrow movements, and smiles). For children similar results were found but did not appear as uncertain as the adults' were. The children in this experiment were aged 7–8 which is younger than the children in Rowland's study [11], where the age was around 10 years (see next section). Age matters: Krahmer and Swerts suggest that young children do not signal uncertainty in the way adults do because they do care less about self-presentation than adults. Our study is about adults only.

6.2.2 Linguistic Pointers to Uncertainty

Knowledge questions, like the ones mentioned above, can be seen as 'testing questions' where the focus may not be on revealing the truth but rather on exposing ignorance and thus adding pressure on the speaker, making him or her nervous and uncertain [1]. Since a common perception about mathematical propositions is that they are either right or wrong, Rowland analyses transcripts of interviews with children focused on mathematical tasks and looks at the children's use of language to shield themselves against accusation of error [11]. According to Rowland, children tend to use a certain category of words (called hedges) which are associated with uncertainty. These hedges are further divided in different types:

- Shield

  o Plausibility shield (*I think, maybe, probably*);
  o Attribution shield (*according to, says . . .*).

- Approximators

  o Rounders (*about, around, approximately*);
  o Adaptors (*a little bit, somewhat, fairly*).

While some hedges are obvious shields to protect against 'failure', others are more elusive and require some contextual information. For example, the word 'about' may be a shield when used in combination with a number (e.g. 'there are about 150,000 people in Enschede') but is no such thing when used in a sentence like 'the story is about a small boy'.

Another research which looks at the use of hedges is that of Bhatt et al. [3]. In their research they study how students hedge and express affect when interacting with both humans and computer systems. It was found that the students hedge and apologize to human tutors often, but very rarely to computer tutors. Another important result of their research is that hedging is not a clear indicator of student uncertainty or misunderstanding, but rather connected to issues of conversational flow and politeness.

6.2.3 Prosodic Markers of Uncertainty

Prosody is important because a speaker can communicate different meanings not extractable from lexical cues by giving acoustic 'instructions' to the listener how to interpret the speech. A good example is the increasing pitch (high F_0) at the end of a question. By using this kind of intonation the speaker draws attention to his question. Other theories include the speaker taking a humble stance by imitating a younger person (with higher F_0 and formants) since s/he is actually asking a favour to the listener (answering his question) [10, p. 277].

In their research Liscombe et al. investigate the role of affect (student certainty) in spoken tutorial systems and whether it is automatically detectable by using prosody [8]. They discovered that tutors respond differently to uncertain students than to certain ones. Experiments with Intelligent Tutorial Systems (ITS) indicate that it is also possible to automatically detect student uncertainty and utilize that knowledge for improvement of these ITS's, making them more humanlike. During their research they not only looked at the current (speaker) turn but also compared this turn with the dialogue history. Among the features analyzed were mean, minimum, maximum and standard deviation statistics of F_0 and the intensity, voiced frames ratios, turn duration and relative positions where certain events occurred.

6.3 Problem Statement

Since much of the research above limits itself to the answering of trivia questions or short answers some question marks can be placed at the usefulness of the results in a broader/different context. Many applications using automatic recognition of the degree of certainty of a person with respect to what s/he is saying might require different input than 'simple' question/answer-pairs. Since the experiments as described above needed relatively short answers (a few words) in order to get a standardized intonation [9] one could wonder what the effects will be on longer utterances like normal dialogues, statements or presentations. Also, a rising intonation (a sign of uncertainty when answering a question) then can also be meant as a question itself (so how to differentiate between the two?) and the latency before an utterance may be irrelevant since the (potentially) uncertain utterance might be encapsulated in other utterances from the same speaker. Nonetheless, these short utterances derived from question answering sessions make it possible to research prosodic features of speech which may be correlated with (un)certainty.

Can prosodic features be used to automatically assess the degree of (un)certainty in a normal spoken dialogue? And which features, if any, qualify best as prosodic markers of this (un)certainty?

From previous research we had already seen that certain features (intonation, latency) can be used to assess the degree of (un)certainty in (short) answers to questions. While the applicability of these features on utterance derived from normal dialogue may be a bit more complex, they are still expected to be valuable indicators. Uncertain utterances will probably have a rising intonation due to the questionable nature of these utterances ('Maybe we can make a green remote?'). Also, common sense would correlate uncertain utterances with longer pauses (latencies) between words.

Besides intonation and latency (or gaps between words in case of longer utterances) We can imagine intensity (softer, less conviction in case of uncertainty) and the speed of talking to be a factor to identify uncertainty. In both cases some way of comparing it to a mean value for these features will be needed, though, since it would not be possible to state whether the utterance has a below/above average value for intensity or speed.

6.4 Data Selection

In order to be able to perform prosodic analysis and reach valid conclusions, it seemed logical to use an existing corpus which had already been annotated. The AMI Corpus, which we addressed during the preliminary phase of this project, not only had many hours of high quality voice recordings but also annotations on different levels (hand made speech transcriptions, time aligned words, dialogue acts) which could be used for this research.

6.4.1 Selection of Meetings

After reviewing the available annotation data for the AMI Corpus [2] a choice had to be made as to which sets were to be analyzed. Since the ES, IS and TS sets were the only ones with complete coverage of the words and dialogue acts annotations and the existence of these annotations was considered essential, these three sets were chosen. As can be seen in Table 6.1, the total dataset now existed of 552 audio files with a total duration of about 280 h.

A disadvantage of the corpus used is the lack of sufficient stance annotations needed for the identification of uncertainty in speech. Since there was no reliable and efficient way to mark uncertain utterances, it was decided to use lexical elements (hedges) to identify utterances which would have a high probability of being uncertain. We split the dialogue acts into three classes: *uncertain* (that contain uncertainty hedges), *certain* (that contain certain hedges), and *neutral* (that do not contain any hedges).

In Table 6.2 an overview of indicators used can be seen. These groups of words are derived from previous studies as performed by Rowland [11] and Bhatt et al. [3]. This approach raises some questions. In their study Bhatt et al. already disputed hedges being only indicators for uncertainty, mentioning they could also be used for politeness strategies [3]. To make sure the assumption made was valid 25 random dialogue acts, marked as uncertain during this research, were ranked on a five point scale ranging from certain to uncertain: certain – probably certain – undecided – probably uncertain – uncertain. 80% of the utterances were scored as either uncertain or probably uncertain.

6.4.2 Data Preparation and Selection

In preparing the AMI data to run through PRAAT, certain errors in the data were found (missing end or begin times of words). Since the Dialogue Act tiers are based

Table 6.1 Overview of selected audio files

	Groups	Meetings	Files	Duration
ES	15	60	240	118:52:35
IS	10	40	152	93:05:28
TS	10	40	160	92:54:05
Total	35	140	552	278:01:50

Table 6.2 Overview of
hedges for uncertainty and
words indicating certainty

Uncertainty	Certainty
According (to)	Absolutely
Approximately	Certainly
Around	Clearly
Fairly	Definitely
Maybe	(In) fact
Perhaps	Must
Possible	Obviously
Possibly	(Of) course
Probable	Positively
Probably	Surely
Somewhat	Undeniably
(I) think	Undoubtedly
Usually	

Table 6.3 Overview of converted words and dialogue acts

	Words			Dialogue acts		
Series	Valid	Invalid	Invalid %	Valid	Invalid	Invalid %
ES	351.615	42	0.01%	47.251	35	0.07%
IS	198.968	14	0.01%	26.909	14	0.05%
TS	283.208	695	0.24%	42.394	419	0.98%
Tot	833.791	751	0.09%	116.554	468	0.40%

on the word tiers therefore several Dialogue Act intervals had missing start and/or end times also and had to be discarded. In Table 6.3 the total amount of valid and invalid items can be seen. Since the percentage of these incorrectly annotated words and dialogue acts was very low it was decided to simply discard them from the dataset instead of trying to figure out the correct data (if possible at all).

PRAAT was used for the prosodic analysis [4]. First a selection of the relevant prosodic features which had to be measured had to be made.

For each category of the prosodic properties mentioned in Section 6.2.3, several attributes were chosen and implemented in PRAAT. Beside these prosodic attributes some lexical attributes (like amount of words, the presence of '*yeah (, but)*', '*okay*') were added as well. In total 76 attributes were chosen for the analysis, of which 67 were prosodic.

The amount of dialogue acts including hedges consists of only 7.26% of the total (7.317 dialogue acts of a total of 100.799), which means that simply classifying each dialogue act as certain gives a score of about 93%. By balancing the dataset the script will take 4.819 random other dialogue acts and combine them with the ones containing hedges to form a new dataset.

6.4.3 Statistical Analysis

Since the dataset preparation script in phase 4 has been designed in such a way that different datasets can be created on the fly, it is easy to compare different prosodic features of different classes. In phase 3 of the research, the prosodic analysis, the presence of several lexical markers or indicators was also checked. Among these markers were the hedges as mentioned before, the group of words (supposedly) indicating certainty, *yeah* and *okay*.

6.5 Experimentation

During the following experiments all datasets were levelled on a 50/50 basis so each 'group' was equally represented. As a result, the baseline (computed with the ZeroR classifier) of all datasets is about 50%. Next, the datasets were classified with the J48 (tree) and NaiveBayes (NB) classifiers. Each classifier was evaluated for accuracy using 10-fold cross-validation. We used the implementation in the Weka toolkit [13]. To determine the key attributes being used for this classification, the input data was also evaluated using the InfoGain attribute evaluator in combination with a Ranker search method.

6.5.1 Hedges –vs– No Hedges

First the dataset with the hedges was analyzed. Out of all 100.799 dialogue acts analyzed with PRAAT in phase 3 only 7.317 contained one or more hedges (see also Table 6.4). These instances were complemented with the same (random) amount of dialogue acts containing no hedges. Based on previous research it was expected that several prosodic features would be good indicators for uncertainty in speech. Among these features were a rising pitch, a declining intensity and a slower rate of speech (more pauses and/or longer average word-length).

In Table 6.5 the results of the analysis can be seen. Two classifiers were used (J48 and NaiveBayes); for each the improvement over the baseline (IOB) is included

Table 6.4 Properties of dataset uncertain hedges –vs– no hedges

Class	Instances
No hedges	7.317 (dropped 85.502)
Uncertain hedges	7.317

Table 6.5 Classification performance of hedges –vs– no hedges including improvement over baseline (IOB)

Baseline (ZeroR)	49.98%			
Features	J48	IOB	NB	IOB
Lexical features (LF)	74.67%	24.69%	71.27%	21.29%
Spectrum related features (SF)	67.70%	17.72%	64.59%	14.61%
Pitch related features (PF)	68.27%	18.29%	67.04%	17.06%
Intensity related features (IF)	63.61%	13.63%	61.20%	11.22%
Formant related features (FF)	66.80%	16.82%	67.97%	17.99%
All prosodic features	66.05%	16.07%	68.46%	18.48%
All features	71.14%	21.16%	69.96%	19.98%
Average Improvement		18.34%		17.23%

in the table. As anticipated, the baseline is about 50% correct classifications. Two striking results are the overall improvement over the baseline score (with an average increase of about 17–18% based on which classifier has been used) and the high performance on the lexical features alone. The evaluation of the (key) attributes show the importance of attributes related to the length of the dialogue act.

The first 8 attributes, headed by the amount of words (da_words) in the DA, are all related to the DA length, either indicating time or the amount of (voiced) frames or bins. Since the utterances in the corpus have been marked (un)certain by using hedges, this is not very surprising: hedges are normally part of (longer) sentences. As a result, the length of a dialogue act (shown by a number of attributes) is a good indicator since short dialogue acts are often marked certain.

After the attributes indicating length in some way the type of dialogue act is also important, taking 9th place in the attribute ranking. Apparently the type of DA as annotated by the members of the AMI Project has some relation to uncertainty. More about the distribution of hedges over dialogue acts can be seen in Section 6.3. Next in the attribute ranking are several formant attributes headed by the minimum F2, maximum F1 and maximum F2. After several other formant attributes the standard deviation for the intensity during the second half of the DA (intensity2_sd), the spectrum band energy (spectrum _band_energy) and the voiced frame ration during the second half (pitch2_voiced_fr_ratio) and the total DA (pitch_voiced_fr_ratio) seem to be good indicators for uncertainty. When classifying the dataset with the J48 classifier and using only the formants' minimum and maximum values the performance result is 67.3%, even higher than when using all formant attributes. Classification based on the voiced frame ratios only gives a performance of 59.8%.

6.5.2 Uncertain Hedges –vs– Certain Hedges

Similar to the previous dataset where dialogue acts with hedges were compared to dialogue acts without these lexical markers, another set was created which contained all dialogue acts with words which should indicate certainty and compared to a similar sized group of hedged dialogue acts. As can be seen in Table 6.6, the size of this dataset was significantly smaller.

In contrast with the expectations mentioned above, the actual results show a lower performance of the classifiers with an average improvement of about 5%. Once again the lexical features score best, although the gap is smaller (Table 6.7).

This time the attribute ranking shows the type of DA (da_type) being the most predictive attribute, followed by some length-related attributes. The first prosodic

Table 6.6 Properties of dataset hedges –vs– anti-hedges	Class	Instances
	has_ hedge[1] = Hedges	663 (dropped 6.654)
	has_ hedge[2] = Anti-Hedges	663

Table 6.7 Classification performance of hedges –vs– anti-hedges including improvement over baseline (IOB)

Baseline (ZeroR)	49.77%			
Features	J48	IOB	NB	IOB
Lexical features (LF)	58.30%	8.52%	57.77%	7.99%
Spectrum related features (SF)	55.66%	5.88%	50.38%	0.60%
Pitch related features (PF)	55.13%	5.35%	52.26%	2.49%
Intensity related features (IF)	53.09%	3.32%	54.45%	4.68%
Formant related features (FF)	51.58%	1.81%	55.13%	5.35%
All prosodic features	55.28%	5.51%	54.90%	5.13%
All features	56.41%	6.64%	55.51%	5.73%
Average Improvement		5.29%		4.57%

feature is the mean F4 (6th place), followed by the minimum intensity (9th) and minimum pitch (12th). In contrast to the previous dataset where the formants played an important role, for this dataset the pitch values (mainly of the 2nd half of the DA) seem to be a better indicator for uncertainty.

Table 6.8 Distribution of (uncertain) dialogue acts

Dialogue acts (ID)	Total dialogue acts	Percentage of total DA's	Hedges	Percentage of hedges	Percentage of dialogue act
Minor	30.816	30.6%	670	9.2%	2.2%
Backchannel (1)	10.655	10.6%	33	0.5%	0.3%
Stall (2)	6.983	6.9%	82	1.1%	1.2%
Fragment (3)	13.178	13.1%	555	7.6%	4.2%
Task	56.438	56.0%	6.094	83.3%	10.8%
Inform (4)	29.841	29.6%	2.456	33.6%	8.2%
Suggest (6)	8.610	8.5%	1.645	22.5%	19.1%
Assess (9)	17.987	17.8%	1.993	27.2%	11.1%
Elicit	6.557	6.5%	396	5.4%	6.0%
Elicit-inform (5)	3.743	3.7%	125	1.7%	3.3%
Elicit-offer-or-suggest (8)	640	0.6%	45	0.6%	7.0%
Elicit-assessment (11)	2.016	2.0%	225	3.1%	11.2%
Elicit-comment-understanding (13)	158	0.2%	1	0.0%	0.6%
Other	6.988	6.9%	157	2.1%	2.2%
Offer (7)	1.370	1.4%	80	1.1%	5.8%
Comment-about-understanding (12)	1.942	1.9%	16	0.2%	0.8%
Be-positive (14)	1.856	1.8%	40	0.5%	2.2%
Be-negative (15)	84	0.1%	3	0.0%	3.6%
Other (16)	1.736	1.7%	18	0.2%	1.0%
Total	100.799	100.0%	7.317	100.0%	5.5%

6.5.3 Distribution of Hedges Over Dialogue Acts

To see whether uncertain utterances occur more in particular dialogue acts the distribution of dialogue acts marked uncertain over the different dialogue act classes has been looked into, the results of which can be seen in Table 6.8. For comparison, the distribution of all dialogue acts has been included as well.

As can be seen in Table 6.8, most dialogue acts are task oriented or minor (56 and 31% respectively). We can also notice that most dialogue acts marked as uncertain (by containing hedges) belong to the task-category.

The class of minor acts contains significantly fewer hedges than the class of elicit acts ($\chi^2(df = 1) = 311.45$; $p < 0.001$) and the class of elicits contains significantly fewer hedges than the class of task acts ($\chi^2(df = 1) = 143.93$; $p < 0.001$).

6.6 Conclusions

Based on the results described in the previous paragraphs, with classification performance increases of up to more than 20%, it is feasible to conclude that the degree of (un)certainty in spoken dialogues can be assessed automatically. When looking at the features which qualify best as prosodic markers of uncertainty the textual features obviously score best. Due to the nature of the uncertain utterances (being based on hedges which most often require some sort of sentence) this result might be of no surprise. There also seems to be a connection between the type of dialogue acts (as annotated by members of the AMI Project) and the degree of uncertainty since the presence of uncertain utterances in several dialogue act types is clearly above average. A relatively high percentage of uncertain dialogue acts are suggestions or assessments. Whether these dialogue acts are really uncertain or whether politeness strategies play a role here is hard to establish.

Another interesting point are the results on which prosodic markers qualify best. It was predicted that a rising intonation, longer pauses (latencies) and a decreasing intensity would be good indicators for uncertainty. Based on the attribute evaluation of the different datasets these theories seem to be supported, showing important roles for the pitch and intensity features. Especially with the dataset 'Hedges –vs– No Hedges' the minimum and maximum values of the formants are good prosodic markers as well.

Even though the results seem straightforward, with impressive classifier improvements over the baseline performances, several questions still remain.

In the current research the feature extraction was based on previous research and the possibilities of PRAAT. While a broad range of features have been researched, it could very well be certain additional features might be promising as well. Another improvement could be using custom settings in PRAAT. For now all settings have been kept on default but it is known that, for optimal results, different settings should be used for men and women, for example. Additional difficulty would be to either automatically detect the gender of a speaker and adapt the settings accordingly, or manually set gender-values for all 500+ files.

For future research on this topic it would be advisable to have a clear understanding of what the 'uncertainty' being researched is and how it can be measured. Having that information should provide a basis for reliable annotations, with which further research can be done.

Further research in hedges and/or other lexical markers as indicators for uncertainty looks promising. The results of combined feature sets have already shown the best results and expanding those features with other indicators (also visual ones) will probably give the best results in the end (although not all types of information will be available in all situations).

Acknowledgments This work is supported by the European IST Programme Project FP6-033812. This article only reflects the authors' views and funding agencies are not liable for any use that may be made of the information contained herein.

References

1. Ainley, J. 1988. Perceptions of teachers' questioning styles. In *12th International Conference for the Psychology of Mathematics Education*, 99–99, Vezprém, Hungary.
2. Carletta, J. C. 2007. Unleashing the killer corpus: Experiences in creating the multi-everything AMI meeting corpus. *Language Resources and Evaluation* 41(2):181–190.
3. Bhatt, K., M. Evens, and S. Argamon. 2004. Hedged responses and expressions of affect in human/human and human/computer tutorial interactions. In *26th Annual Meeting of the Cognitive Science Society*, Chicago, Illinois.
4. Boersma, P., and Weenink, D. 2008. Praat: Doing phonetics by computer. Version 5.0.06 [Software]. Available: http://www.praat.org/.
5. Brennan, S. E., and M. Williams. 1995. The feeling of another's knowing: Prosody and filled pauses as clues to listeners about the metacognitive states of speakers. *Journal of Memory and Language* 34:383–398.
6. Hart, J. T. 1965. Memory and the feeling-of-knowing experience. *Journal of Educational Psychology* 56:208–216.
7. Krahmer, E., and Swerts, M. 2005. How children and adults signal and detect uncertainty in audovisual speech. *Language and Speech* 48(1):29–54.
8. Liscombe, J., J. Hirschberg, and J. J. Venditti. 2005. Detecting certainness in spoken tutorial dialogues. In *9th European Conference on Speech Communication and Technology*, 1837–1840. Lisbon.
9. Ozuru, Y., and W. Hirst. 2006. Surface features of utterances, credibility judgments, and memory. *Memory & Cognition* 34:1512–1526.
10. Rietveld, A. C. M., and V. J. Van Heuven. 1997. *Algemene Fonetiek*. Bussum: Uitgeverij Coutinho.
11. Rowland, T. 1995. Hedges in mathematics talk: Linguistic pointers to uncertainty. *Educational Studies in Mathematics* 29:327–353.
12. Smith, V. L., and H. H. Clark. 1993. On the course of answering questions. *Journal of Memory and Language* 32:25–38.
13. Witten, I. H., and E. Frank. 2005. *Data Mining: Practical Machine Learning Tools and Techniques*, 2nd ed. San Francisco: Morgan Kaufmann.

Chapter 7
Metaphors and Metaphor-Like Processes Across Languages: Notes on English and Italian Language of Economics

Maria Teresa Musacchio

7.1 Introduction

The debate on the nature and understanding of metaphorical meaning is an important one, especially in the context of changes in theories of physical sciences [3] and more generally in the so-called paradigm shift in these sciences [13]. Major changes in science have invariably involved researchers across linguistic boundaries working together to create, for example, quantum theory, nuclear fission, polymer compounds, nucleic acids, and many more of these iconic changes [1]. The list of Nobel Laureates in physics and chemistry over the last 100 years clearly shows the multilingual nature of changes in science; see also, for example, Pais' [18] account of the developments in nuclear and particle physics between 1895 and 1983 which is interspersed with notes about the competence in English of major figures like Enrico Fermi and other researcher/migrants to the USA. The multi-lingual interaction in life sciences has been a subject of textual analysis and metaphors across languages appear to play a key role [24, 25]. So, one can argue that metaphors play a key role in the development of sciences and that the process of discovery, and the subsequent institutionalisation of scientific concepts, procedures and instruments, involves a community of scientists who have different first languages. There are two different types used in science: exegetical or pedagogic metaphors are regarded as playing a role in teaching or explaining theories, while constitutive metaphors are – at least for some time – considered as an irreplaceable part of a scientific theory [3, pp. 359–360]. It has been noted that there are differences between a metaphor and a metaphor-like process and that long after the process of exploring the potential similarities or analogies between the source and target domains has ended, a metaphor can remain essential to a theory to the point that the metaphor-like process can be regarded as hardly ever completed [14, pp. 414–415].

In this paper, I look at these two key types of metaphors in a social science – economics – that has a substantial mathematical framework like physical sciences

M.T. Musacchio (✉)
Dipartimento di Lingue e Letterature AngloGermaniche e Slave, Università di Padova,
Via Beldomandi 1, 35137 Padova, Italy
e-mail: mt.musacchio@unipd.it

K. Ahmad (ed.), *Affective Computing and Sentiment Analysis*, Text, Speech
and Language Technology 45, DOI 10.1007/978-94-007-1757-2_7,
© Springer Science+Business Media B.V. 2011

and yet – being a social science – draws more, perhaps, from everyday experience of economic transactions [2]. World trade, expedited by fast means of communications and logistics, and regulated by international organisations like the IMF and WTO, has exacerbated the demands for having documents written in the special language of economics translated extensively. It has been noted that economists use both pedagogic and constitutive or conceptual metaphors (the work of Henderson [8–10], Lakoff and Johnson [15], and McCloskey [16] is particularly relevant here); an innovation in economics is that of *heuristic* metaphors – for 'catalysing thinking' or 'propelling thoughts' (see, for instance [12, 14]). Typical examples of constitutive metaphors in economics are the *production function* and *market equilibrium* originating from the mechanistic approach which has been the dominant paradigm in neoclassical economics [4, p. 66], whilst the *circular flow diagram* exemplifies pedagogic metaphors [4, p. 61]. With reference to these two different metaphor types, my corpus-based analysis shows that constitutive metaphors dominate reports and speeches (of key stakeholders) whilst pedagogic metaphors are more frequent in opinion-opposite-editorial (op-ed) text and magazine articles; the divide in the usage relates to the classical division of texts – informative (reports/speeches) and imaginative (op-ed and magazine articles) in corpus-based studies (BNC, Brown, Lancaster/Oslo Bergen).

Despite recent claims of attention to metaphor in (con)text and metaphor function, researchers in translation studies [20, 21, 23] have partly overlooked that consideration of text type – at least the usual tripartite classification in academic, instructive and popular-science texts – is essential when discussing any feature of science and its language, including metaphor in economics. It has been known that popular economics texts differ from other text types as they contain explanations or definitions of concepts [10], their metaphors are much more extended than in specialized texts, and they exhibit a simpler syntax than academic or institutional texts [9]. When referring these remarks specifically to metaphor-building, one could infer that in popular economics texts as – more generally – in popular science, some metaphors are culture-specific since they arise 'both from cognitive, bodily constraints and from shared experience' [6, p. 181]. Cross-linguistic research to investigate the possibility that metaphors are not language-specific has shown that at least some of them are shared. However, cases were detected where no complete consistency could be found. In particular, studies of metaphors in economics texts in English, French and Dutch have shown differences that were ascribed to cultural factors [5, pp. 1232–1233].

I have investigated metaphors in economics by analysing three text types – economic reports, speeches written to be read and magazine or newspaper articles – and contrasted the use of metaphorical processes in articles originally written in Italian and articles translated from English into Italian. This corpus-based inquiry is meant to be both intra- and interlingual. It is intralingual as it covers three text types which – unlike academic papers – are still quite commonly written in Italian, i.e. reports as sources of economic information, speeches by two stakeholders in the field, the former and the current Governors of the Bank of Italy, and magazine/newspaper articles as instances of popular economics [10, 11]. My inquiry is also interlingual, or cross-linguistic, since it compares and contrasts use of metaphors in original and translated Italian articles on economics to highlight possible influences of the English sources on Italian translations and alternative formulations of metaphors in the two languages.

7.2 Corpus and Method

7.2.1 Corpus

Data were collected through analysis of a corpus consisting of four sub-corpora: (a) a set of economic reports published by the Bank of Italy and ISTAT, the Italian National Institute of Statistics, (b) speeches delivered by the former and the current Governors of the Bank of Italy, Antonio Fazio and Mario Draghi; (c) articles from the Italian economic and financial daily *Sole 24 Ore*, from the economic pages of the national Italian dailies *Corriere della Sera* and *La Stampa*, and from the Italian business weekly *Economy*; (d) articles from The *Economist* and *The World In* series published by The Economist Publications, translated into Italian and published first by *Economy* and later in special supplements of the Italian daily *La Stampa*, and articles on the international economy published in the *Financial Times* and *Project Syndicate* and translated by the *Sole 24 Ore*.

My three million plus word corpus comprising 4,595 texts which were published between 1995 and 2008, is both comparable – a, b, c – and parallel (d), since it presupposes English source texts – to be examined intra- and inter-lingually. All corpus components are authoritative texts as they include sources of economic information in Italy – Bank of Italy and ISTAT –, texts by leading economists who write on current economic topics (speeches by two Governors of the Bank of Italy), and influential newspapers and magazines – *Sole 24 Ore, Corriere della Sera, La Stampa* and translations from *The Economist* published in *Economy* and *La Stampa*. Corpus components, size, and text types are summarized in Table 7.1 with indication of the reasons why they are regarded as authoritative.

7.2.2 Method

For the 'fear' of missing out on a number of metaphorical expressions, corpus-based studies often resort to making a decision as to what lexemes to focus on [5] or restrict analysis to one feature – metaphor to do with *inflation* [7]) – or to one text type [22]. Without any attempt at exhaustiveness, a different method is used in this paper. The corpus was analysed to look at constitutive metaphors, especially the spatial or orientational metaphors – *up* and *down* metaphors – à la Lakoff and Johnson [15, pp. 14–19] – and 'organicist' metaphors considering economies as organisms and related to recurring changes or variation – business cycle metaphors. I also looked at pedagogic metaphors which include war metaphors – following White's *business*

Table 7.1 Corpus components grouped into the four sub-corpora of economic reports, governors' speeches, original Italian articles and translated Italian articles with indication of their corresponding authority. The corpus includes virtually all the Economist articles published in translation by Economy as long as it had exclusive copyright for Italy and 45 speeches of the current Governor of the Bank of Italy, Mario Draghi, up to August 2009. These are matched by 45 speeches by the former Governor, Antonio Fazio, which were selected starting from the last one in 2005 and proceeding backwards – again to ensure corpus balance

Text type	Source	Tokens	No. of texts	Authority
Annual reports	Bank of Italy, ISTAT	689,153	7	Sources of economic information
Speeches	Governors of the Bank of Italy	481,435	90	Leading economic stakeholders
Original Italian articles	*Sole 24 Ore, Corriere della Sera, Economy, La Stampa*	1,888,116	3406	Influential newspapers/magazines
Translated Italian articles	*Economy, La Stampa Suppl., Sole 24 Ore*	102,303	92	Influential newspapers/magazines
Total		3,161,007	4595	

is war notion [26, p. 134] modelled on Lakoff and Johnson's *argument is war*, [15, p. 4], and weather related metaphors as notions of *heat* (cooling and heating) and *entropy* (lethargy) which are used across scientific disciplines.

When I argue that I have looked for the constitutive (or pedagogic) metaphor, I mean that I look at frequency counts, concordance and informativeness of single and compound tokens in each of the 4,595 texts which in my opinion have been used as a metaphor for one type or the other. I have noted that 16 tokens were used in a metaphorical sense specifically as constitutive metaphors and 10 other tokens were used as pedagogic metaphors.

7.3 Analysis

7.3.1 Constitutive Metaphors

The highest number of occurrences in my four subcorpora includes words or terms describing change, that is variation taking the form of a *pattern* or *trend*. There are a number of semi-technical terms[1] drawn from physics – *dinamica/dinamiche* (trend, from *dinamica* dynamics) and *flessione/i* (flexion, bending; metaphorically drop, downturn) – and a general language word, *andamento* (pattern, trend). Key metaphors for *increase* and *decrease* (Lakoff and Johnson's UP and DOWN

---

[1] With reference to economics, Henderson [10, p. 171] defines a semi-technical term as a term that is "metaphorically derived but has a precise meaning, which is, in context, understood, but choice can be exercised with respect to its use". As examples of semi-technical terms in economics Henderson gives *impact* and *bid*.

metaphors) – *crescita* (growth, based on the idea of the economy as a living thing), *rialzo/i* (lit. a rise, but also elevation, wedge) and *ribasso/i* (reduction, drop, decline, but also discount, rebate).

The use of *dinamica* as a semi-technical term, in many different fields, dates back to 1862, *flessione* was first used with the meaning of 'progressive reduction' in 1935, while *andamento* is the abstract noun from the verb *andare* (to go) and first appeared in the early fourteenth century. *Rialzo* was first used with reference to the stock exchange in 1797 and to prices in 1872; *ribasso* was used to signify a decrease in prices or values appeared in 1745, while it was assigned to variation in the stock exchange in 1867.[2] Most of these metaphorical terms developed in the early days of the science of economics when the sources of terminology were mainly Italian and French.

I have found that *dinamica* collocates with terms such as investment, production, prices and consumption and is also used to refer to the trend of the business cycle (*dinamica congiunturale*). Further, as they only describe a pattern metaphorically, *dinamica* and *andamento* collocate with adjectives such as *sostenuto* (sustained), *moderato* (moderate), *favorevole* (favourable, positive), and *sfavorevole* (unfavourable, negative) that attach a degree of 'sentiment' to data and thus guide readers in the interpretation of data (see, for example, Table 7.2).

Rialzo and *ribasso* are most frequent in original Italian articles. Furthermore, a successful metaphor engenders a metaphor-like process that gives rise to compounds (*revisione al rialzo/al ribasso*, literally translated: upward/downward adjustment) and derivatives (the verb *rialzare* and the adjective *rialzista*). In newspaper and magazine articles metaphors appear to be freely mixed: *revisione al rialzo* is followed by the metaphorical *andamento* and *rialzare* – in the collocation *rialzare i tassi* (to increase rates) is used to pour oil on the troubled waters of prices and recovery or, in Italian, *per gettare acqua sul fuoco dei prezzi e della ripresa*, which literally translates as 'to throw water on the fire of prices and recovery' (see Table 7.3).

I have noticed that comparing the relative frequency of metaphors signalling economic change in original Italian texts with those in translated articles shows that originals use all metaphors more frequently apart from *crescita* (growth), which is not as common, and *ribasso*, which is equally frequent in all texts. Like growth

Table 7.2 A concordance of *dinamica, andamento, flessione* and *crescita* [sub-corpus: reports and speeches]

LH co(n)text	Term	RH co(n)text
eccezione delle costruzioni nel 2007, la	dinamica	della produttività del lavoro nel
eccezionali proventi connessi con il positivo	andamento	dei mercati finanziari
capitale bancario hanno provocato una	flessione	dell'attività economica
dell'intero saldo finanziario del settore. La	crescita	si è attenuata nella seconda metà

[2] All information on etymology is taken from Manlio Cortelazzo and Paolo Zolli's *DELI – Dizionario Etimologico della Lingua Italiana*, 2nd revised edition by Manlio Cortelazzo and Michele A. Cortelazzo published by Zanichelli in Bologna in 1999.

Table 7.3 A concordance of *rialzo, ribasso* and/or their derivatives or compounds [sub-corpus: Italian originals]

LH co(n)text	Term	RH co(n)text
ha indotto fin da subito una significativa	revisione al rialzo	delle previsioni sull'andamento
deve ancora mostrare un sostenuto	trend rialzista",	prosegue il rapporto.
si troverà ben presto nella necessità di	rialzare	i tassi per gettare acqua sul fuoco
La	revisione al ribasso	è ampia per tutti i grandi paesi di

in English, *crescita* is both a metaphorical term to indicate the wealth of a nation measured as a rise in GDP, and a semi-technical one used as a synonym of increase and aumento. As such it appears to straddle the fuzzy boundaries between organicist constitutive metaphors [26, p. 132] and biological pedagogic ones [10, p. 170]. Metaphors taken from physics (*dinamica* and *flessione*) are most frequent in reports which – as sources of economic information – reflect formal use in economics. Orientational metaphors (*rialzo* and *ribasso*) are equally common in original Italian texts. The swinging frequencies of most of these metaphors in Italian originals suggest that a factor contributing to metaphor variation may be text type. (See Table 7.4 for details).

All metaphors described above are often found in the context of the business cycle – a key concept in economics termed in Italian *ciclo economico*, when it refers to change over the medium or long term, and *congiuntura* (literally joint; but also situation, circumstance) in the case of short-term variation. Its stages are described by means of mechanistic and organicist metaphors drawn respectively from physics (UP and DOWN metaphors cited by Lakoff and Johnson), and health and medicine: *espansione* (expansion), *contrazione* (contraction), *rallentamento* (slowdown); *crisi* (crisis), *depressione* (depression), *ripresa* (recovery). The same metaphors are used in Italian and English and *boom* is also used as a loan word from English meaning (rapid) expansion. The metaphorical terminology of business cycles in Italian has two main origins. Some terms such as *ciclo, espansione* and *rallentamento* are still regarded as extensions of meanings either in the general language – *ciclo*, c. 1575 – or in physics – *espansione* (c. 1630); *rallentamento* (fourteenth century). Other terms were entered in Italian dictionaries in the first half of the twentieth century: *crisi* (1925), *contrazione* and *boom* (1931), *depressione* (1940), *recessione* and *congiuntura* (1942), *ripresa* (1946). This is hardly

Table 7.4 The distribution of metaphors of change in original Italian text types contrasted with use in translated articles

Metaphor	Semi-technical term	Reports	Speeches	Original articles
UP	crescita	0.75	0.8	0.5
DOWN	ribasso/i	1.0	1.0	1.0
Physics: trend	dinamica/che	10	4.3	1.4
	rialzo/i	2	2	2
	flessione/i	15	10	2.5
Physics: pattern	andamento/i	5	12.5	5

Table 7.5 Metaphors of (stages in) business cycles in original Italian texts compared with translations

Metaphor	Term	Reports	Speeches	Original articles	Translated articles
DOWN organicist	depressione	0.005	0.15	0.1	YES
	recessione	0.2	0.2	0.2	YES
	ripresa	0.3	0.8	0.7	YES
Physics: Mechanistic	rallentamento	1.3	0.7	0.3	
	boom		0.01	0.25	
DOWN	contrazione	3	1	0.2	
Physics: Mechanistic	ciclo economico	1	1	0.5	
UP	espansione	1	2	0.5	
	congiuntura	3	7	3	NO
Organicist	crisi	6	9	6	NO

surprising as the first systematic description of business cycles was Schumpeter's in 1939. Indeed, the Italian *congiuntura* in its economic meaning is a loan translation from German *Konjunktur*. If the method used for 7.4 is applied to metaphorical terms for stages in the business cycles, *congiuntura* and *crisi* emerge as the least frequent metaphors in translated articles. In the more formal text types – reports and speeches – UP and DOWN metaphors such as *espansione* and *contrazione* are preferred, while original articles make extensive use of organicist metaphors such as *crisi, recessione, ripresa* and the mechanistic metaphor *rallentamento*. (See Table 7.5).

7.3.2 Pedagogic Metaphors

Metaphor in economics can further draw on physics to describe changes in business cycles. In this case, too, the two constitutive metaphors, based on mechanistic processes, are represented here by the two stems *accel** – as in *accelerare* (to accelerate), *acceleratore* (accelerator) and so on –, and *decel** (slowdown). Owing to their productivity, they are most frequent in reports and speeches whilst the pedagogic metaphors based on the more common *fren** stem (brak(e)*, curb*) designate economic change when slowdown sets in.

7.3.3 Universal vs. Culture-Specific Metaphors

Among pedagogic metaphors which are more common in argumentative texts I have chosen two sets to highlight differences in usage in original and translated Italian articles. War and conflict as metaphors of economic and business endeavours were found to be more common in translated articles than other subcorpora. The stems

lott* (*lotta/e*, fight(s); *lottare*, to fight), *conflitt** (*conflitto/i*, conflict(s)), and *sfid**
(*sfida/e*, challenge(s); *sfidare*, to challenge) show lowest frequency in economic
reports that mainly deal with macroeconomic topics such as business cycles and
overall trends in economies. The pedagogic metaphors of war and conflict may be
better suited to popular business and financial texts and thus be more frequent in the
articles of this corpus, while the highest frequency of *sfid** in translations points to
greater relevance of this metaphorical sub-domain in English as opposed to *lott** in
Italian.

Finally, analysis of the corpus suggests that typically culture-specific metaphors
in Italian concern the housing market. In economic reports the general language
word *abitazione/i* (dwelling/s) is used or the term for the corresponding market,
mercato immobiliare (housing market) is referred to. In original Italian articles,
however, real estate is metaphorically termed *casa/e* (house, but also home) as an
icon of family in contrast with its economic equivalent, the household. In the language of the press, the building industry and housing market are also metonymically
referred to by one of their basic components, the *mattone* (brick), which is not used
at all in reports and speeches.

7.4 Conclusion

Data analysis in this paper suggests that use of metaphors in economics varies
according to text type and – to a lesser extent – from language to language. In
economic reports, the prevailing constitutive metaphors reflect standard use in economics and their surrounding co(n)texts express sentiment in a highly controlled
fashion, i.e. with frequent hedging and downtoning. Conversely, newspaper and
magazine articles use pedagogic metaphors more frequently – and more freely by
mixing them and adding expressions of much more emotionally charged language
and imagery. Translated texts occupy the middle-ground, as they frequently employ
what are standard constitutive or pedagogic metaphors in the language of economics
but they also try to conform to what is perceived as the norm in the text type, i.e. articles, while at times resorting to culture-specific metaphors that are more widespread
in the source language and domain.

Contrary to linguistic approaches to metaphor such as Partington's [19] which
tend to view metaphors as becoming increasingly standardised into terms, productivity of metaphors detected in the corpus confirms Kuhn's [14] idea that a
metaphor-like process in science continues to play a role until its potential has
largely been exploited and even when the metaphor has lost its currency. In the
present study an attempt has been made to show how a metaphor-like process is
triggered as in the case of *rialzo/revisione al rialzo* and the subsequent mixing
of metaphors. In line with Partington's [19] findings, however, the present study
found low frequency of pedagogic metaphors relating to war, no extended pedagogic metaphors with *flusso* (flow), but frequently mixed metaphors. Analysis also
suggests that some metaphors may be at least partially culture-specific – in line with

results of the study of metaphor in general Italian by Deignan and Potter [5]. The clearest case is that of *casa* and *mattone*, but even the higher frequency of metaphors relating to the weather in the translated texts indicate some difference in metaphor-building in Italian and English economics that could be further investigated.

My analysis of original and translated texts, in a specific domain, e.g. economics, suggests that use of metaphors varies according to text type and – to a lesser extent – from language to language. In economic reports, the prevailing constitutive metaphors reflect standard use in economics and their surrounding co(n)texts express sentiment in a highly controlled fashion, i.e. with frequent hedging and downtoning. Conversely, newspaper and magazine articles use pedagogic metaphors more frequently – and more freely by mixing them and adding expressions of much more emotionally charged language and imagery. Translated texts occupy the middleground, as they frequently employ what are standard constitutive or pedagogic metaphors in the language of economics but they also try to conform to what is perceived as the norm in the text type, i.e. articles, while at times resorting to culture-specific metaphors that are more widespread in the source language and domain.

References

1. Ahmad, K. (2006). Metaphors in the languages of science? In *New Trends in Specialized Discourse Analysis*, Gotti, M. and Bhatia, V. Bern: Peter Lang.
2. Black, M. (1962). *Models and Metaphors: Studies in Language and Philosophy*. Ithaca, NY: Cornell University Press.
3. Boyd, R. (1979). Metaphor and theory change: What is "metaphor" a metaphor for? (pp. 356–408). In: Ortony, A. (ed.) *Metaphor and Thought*. Cambridge: Cambridge University Press.
4. Cosgel M. (1996). Metaphors. Stories and the entrepreneur in economics (pp. 57–76). *History of Political Economy*. 28, 1.
5. Deignan, A. and Potter, L. (2004). A corpus study of metaphors and metonyms in English and Italian (pp. 1231–1352). *Journal of Pragmatics*, 36.
6. Eubanks, P. (1999). Conceptual metaphor as rhetorical response. A reconsideration of metaphor (pp. 171–199). *Written Communication*, 16 (2).
7. Fuertes-Olivera, P. and Pizarro-Sánchez I. (2002). Translation and 'similarity-creating metaphors' in specialised languages (pp. 43–73). *Target* 14 (1).
8. Henderson, W. (1982). Metaphor in economics (pp. 147–157) *Economics*, 18 (4).
9. Henderson, W. (1993). The problem of Edgeworth's style (pp. 200–222). In: Henderson, W., Dudley-Evans, T. and Backhouse, R. (1993). *Economics and Language*. London: Routledge.
10. Henderson,W. (2000). Metaphor, economics and ESP: some comments (pp. 167–173). *English for Specific Purposes*, 19.
11. Hundt M. (1998). Typologien der Wirtschaftssprache: Spekulation oder Notwendigkeit? (pp. 98–115). *Fachsprache*, 20 (3–4).
12. Klamer, A. and Leonard, T.C. (1993). So what's an economic metaphor? (pp. 20–53). In: Mirowski, P. (ed.) *Natural Images in Economic Thought*. Cambridge: Cambridge University Press.
13. Kuhn, T. (1962). *The Structure of Scientific Revolutions*. Chicago: University of Chicago Press.
14. Kuhn, T. (1979). Metaphor in science (pp. 409–419). In: Ortony A. (ed.) *Metaphor and Thought*. Cambridge: Cambridge University Press.

15. Lakoff, G. and Johnson, M. (1980/2003). *Metaphors We Live By*. Chicago: University of Chicago Press.
16. McCloskey D. (1988). *La retorica dell'economia* (Italian translation of *The Rhetoric of Economics*). Torino: Einaudi.
17. McCloskey, D. (1995). Metaphors economists live by. Social Research, Summer, http://findarticles.com/p/articles/mi_m2267/is_n2_v62/ai_17464378/ (last visited on Jan 27, 2011).
18. Pais A. (1986) *Inward Bound*. Oxford, Clarendon Press.
19. Partington, A. (1998). *Patterns and Meanings. Using Corpora in English Language Research and Teaching*. Amsterdam/Philadelphia: John Benjamins.
20. Pisarska, A. (2004). Metaphor and other tropes as translation problems: A linguistic perspective (pp. 520–527). In: Kittel H., Frank, A.P., Greiner, N., Hermans, T., Koller, W., Lambert, J. and Paul, F. (eds) *Übersetzung / Translation / Traduction*. Berlin, Mouton de Gruyter,
21. Schäffner, C. (2004). Metaphor and translation: some implications of a cognitive approach (pp. 1253–1269). *Journal of Pragmatics*, 36.
22. Skorczynska Sznajder H. (2010). A corpus-based evaluation of metaphors in a business English textbook (pp. 30–42). *English for Specific Purposes* 29.
23. Štambuk, A. (1998). Metaphor in scientific communication (pp. 1–7). *Meta* XLIII, 3.
24. Temmerman R. (2000). *Towards New Ways of Terminology Description*. Amsterdam, Benjamins.
25. Temmerman, R. (2002). Metaphorical models and the translator's approach to scientific texts (pp. 211–226). *Linguistica Antverpiensia* 1.
26. White, M. (2003). Metaphor and economics: the case of growth (pp. 131–151). *English for Specific Purposes*, 22.

Chapter 8
The 'Return' and 'Volatility' of Sentiments: An Attempt to Quantify the Behaviour of the Markets?

Khurshid Ahmad

8.1 Introduction

Sentiment analysis is now becoming an established tool for the analysis of financial and commodity markets. The roots of this subject lie in the earlier work (c. 1950s) on content analysis on the one hand and on the other in work on bounded rationality and 'herd behaviour' by Herbert Simon and Daniel Kahnemann. Information extraction and corpus linguistics have been used in extracting the distribution of the so-called affect words and their collocates.

We begin this paper with a mini tutorial on the two key metaphorical terms used in finance studies, especially in the study of changes of prices and that of the value of the indices of a market: *return* and *volatility*. We briefly describe some related work in sentiment analysis. This is followed by a description of a corpus-based study of the variation in the frequency of positive and negative words, as defined in the *Harvard Dictionary of Affect*. An afterword concludes the paper.

8.2 Metaphors of 'Return' and of 'Volatility'

The literature on financial economics that is closely related to the analysis of market sentiment frequently refers to two key terms: *return* and *volatility*. These terms have retained much of their original meaning that is, 'return' broadly refers to the 'act of coming back', as established upon the entry of this word in the English language in the fourteenth century or thereabouts. This word has been adapted, or to put loosely has been used as a metaphor for coming back as in the definition: 'Pecuniary value resulting to one from the exercise of some trade or occupation; gain, profit, or income, in relation to the means by which it is produced'. In financial economics a return, or 'price change quantity' is defined as the logarithm of the ratio of the

K. Ahmad (✉)
School of Computer Science and Statistics, Trinity College, Dublin 2, Ireland
e-mail: kahmad@scss.tcd.ie

K. Ahmad (ed.), *Affective Computing and Sentiment Analysis*, Text, Speech
and Language Technology 45, DOI 10.1007/978-94-007-1757-2_8,
© Springer Science+Business Media B.V. 2011

current and past price (or an index of a stock exchange or any other aggregated index, like Standards & Poor, Financial Times-Stock Exchange Index):

Let p_t be the price today and p_{t-1} the price yesterday, so the return r_t is defined as:

$$r_t = \ln\left(\frac{p_t}{p_{t-1}}\right)$$

The word 'volatility' is much more graphic as it started its journey into the English language from Latin in the seventeenth century: 'The quality, state, or condition of being volatile, in various senses' and the metaphorical use of this word includes references to 'tendency to lightness, levity, or flightiness; lack of steadiness or seriousness'. Benoit Mandelbrot [14] has argued that the rapid rate of change in prices (the *flightiness* in the change) can and should be studied and not eliminated – 'large changes [in prices] tend to be followed by large changes – of either sign – and small changes tend to be followed by small changes'. The term *volatility clustering* is attributed to such clustered changes in prices. Mandelbrot's paper drew upon the behaviour of commodity prices (cotton, wool and so on), but volatility clustering is now used in for almost the whole range of financial instruments (see [17] for an excellent and statistically well-grounded, yet readable, account of this subject).

There are different kinds of measures of volatility, a commonly used version is called *realized* or *historical volatility*. Volatility (v) of a stock price or the value of an index is defined over a trading period n and is the standard deviation of the past returns ($\ln(\frac{p_t}{p_{t-1}})$):

$$v = \sqrt{\frac{1}{n-1}\sum_{k=1}^{n}(r_{t-k} - \bar{r})}$$

where $\bar{r}$ is the average value of past returns in the period n. Econometricians have observed that it is perhaps easier to conduct a statistical analysis of changes in market prices as there is little or no correlation between consecutive price changes; there is a significant correlation in the prices. It has been argued by Robert Engle [9], the 1993 co-winner of the Nobel prize in economics, that '[a]s time goes by, we get more information on these future events and re-value the asset. So at a basic level, financial price volatility is due to the arrival of new information. Volatility clustering is simply clustering of information arrivals. The fact that this is common to so many assets is simply a statement that news is typically clustered in time.' (1993:330).

The term *news arrivals or information arrivals* is defined rather differently in the literature in finance. This standard practice in financial economics is either to use daily counts of news stories as a proxy for information arrival or, in simulations, a random number generator is used for generating the number of news arrivals [5]. This method has been refined to count only those stories that comprise a given (set of) keyword(s) [6, 7], and more recently Tetlock [18] has used affect words in the

General Inquirer lexicon [16] to correlate market movements with change in the frequency of affect words. More of this later.

Let us look at the concepts of return and volatility in the context of *information arrivals* by looking at the recurrence of news items containing the same keywords and, more importantly, on the recurrence of the same metaphorical terms. Consider the use of the term *credit crunch*: According to Wikipedia (2008) the term is defined as 'a sudden reduction in the availability of loans (or "credit") or a sudden increase in the cost of obtaining a loan from the banks.' This term has been around for over 30 years or more, but let us look at the usage of this term since 1981: The *New York Times* archives were searched for the compound term 'credit crunch' and over 400 articles, published between 1981–2008, were found that contained the term. There may have been other articles which the NYT authorities did not or could not include, but the NYT is considered as an authoritative and usually un-biased source of US and international political and economic reporting. The number of stories containing the term appear to have a 5 year cycle, except in the last decade where 'credit crunch' only appeared in large number in 2007; in 2008 there were 54 stories compared to whole of 2007 where there were 77 stories in all. My projected value of the number of stories in 2008 is 156, based on the current average of about 13 a month (Table 8.1).

Let us look at how the 'returns' of the number of stories in Table 8.1 and compute the annual historical volatility purely on the basis of the changes in the numbers of stories in *New York Times*. It has been argued that volatility, computed using return of prices or index values, increases during crises periods, say during the 1929 US Great Depression, the period leading upto the resignation of the US President Nixon in 1974, and the days after the 2001 9/11 attacks (see [17, pp. 191–93]).

We have plotted the number of stories per year, the consecutive year returns, and the volatility for a reporting ('trading'?) period of 5 years. There is a much greater variation in the returns (based on the current and past numbers of news stories containing the term 'credit crunch') when compared to the fluctuations recorded in the actual number of stories. One quantification of such a fluctuation is volatility or the standard deviation of past returns. The volatility increased every 5 years until 2000 – indicating an *increasing* sense of 'crises'. Volatility decreases during calmer periods. In our extremely simple illustration, we note that the volatility decreased during 2001–2005 – a period of massive growth partly fuelled by 'easy credit' – and lastly

Table 8.1 The number of stories per year comprising the term *credit crunch* that appeared in (or are in the archive of) *New York Times*

Year	1981	1982	1983	1984	1985	1986	1987	1988	1989	1990	Decade Total
# Stories	19	6	4	4	8	5	3	4	6	59	118
Year	1991	1992	1993	1994	1995	1996	1997	1998	1999	2000	
# Stories	93	38	24	4	10	3	2	47	3	4	228
Year	2001	2002	2003	2004	2005	2006	2007	2008	2009	2010	
# Stories	10	4	1	2	1	2	225	616	168	59	1088

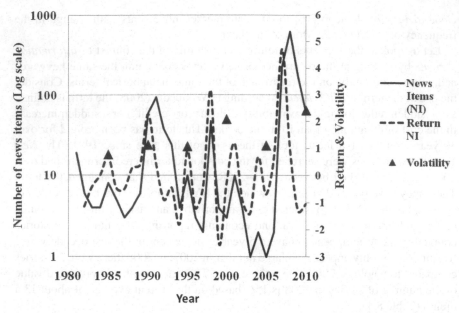

Fig. 8.1 The number of stories are on the left vertical stories, and the percentage change in returns and volatility are on the right

for the 5 years (2006–2010) the volatility has increased dramatically incorporating the period of the very frequent use of the term 'credit crunch' (Figure 8.1):

The above figure shows that a single statistic, volatility, can be used in the quantification of rapid changes. The questions is this: will this statistic throw any light on the changes in market sentiment, based on methods in sentiment analysis that use the frequency of positive and negative affect words as a measure of sentiment?

8.3 The Roots of Computational Sentiment Analysis

Sentiment is defined as 'an opinion or view as to what is right or agreeable' and political scientists and economists have used this word as a technical term. When sentiments are expressed through the faculty of language, we tend to use certain literal and metaphorical words to convey what we believe to be right or agreeable. There are a number of learned papers and reviews in computational sentiment analysis that are available ([11] and references therein).

One of the pioneers of political theory and communications in the early twentieth century, Harold Lasswell [12], has used sentiment to convey the idea of an attitude permeated by feeling rather than the undirected feeling itself. (Adam Smith's original text on economics was entitled *A Theory of Moral Sentiments*.) Namenwirth and Laswell [15] looked at the Republican and Democratic party platforms in two periods 1844–1864 and 1944–1964 to see how the parties were converging and how language was used to express the change. Laswell created a dictionary of affect words (*hope*, *fear*, and so on) and used the frequency counts of these and other words to quantify the convergence.

This approach to analysing contents of political and economic documents – called content analysis – was given considerable fillip in the 1950s and 1960s by Philip Stone of Harvard University who created the so-called General Inquirer System [10, 16] and a large digitised dictionary – the *GI* Dictionary comprising over 8,500 words carefully selected using a criterion developed by the psychologist Charles Osgood including positive/negative words, words to express strength and weakness, and words to describe activity and arousal (Stone's dictionary includes a number of entries used by Harold Laswell; these entries are thus labelled).

Recently, the digitised Harvard *Dictionary of Affect* has been used to 'measure' sentiment in the financial markets. Tetlock [18] has analysed a commentary column in the *Wall Street Journal* using the GI Dictionary and correlated the frequency of affect words with trading volumes of shares in the New York Stock Exchange: He concludes that 'high media pessimism predicts downward pressure on market prices followed by a reversion to fundamentals, and unusually high or low pessimism predicts high market trading volume.' This is amongst the first reported study in financial sentiment analysis that is rooted strongly in econometric analysis (especially through the use of auto-regressive models in the framework of conditional heteroskedasticity) that has analysed the contents of the news in conjunction with the study of information arrival (see also [13]). Tetlock's selection of comment or opinion in a newspaper, classified as imaginative writing rather than the informative news reportage, may raise some methodological questions in text analysis about whether or not opinions can in themselves comprise a representative sample of texts that has been used for analysis (see, for example [2, 4]).

8.4 A Corpus-Based Study of Sentiments, Terminology and Ontology Over Time

We report on some work recently carried on compiling a representative or random sample of texts, in a given domain, that can be used for analysing sentiments. Once the a corpus is compiled we then extract terminology that is used in the domain automatically. Then significant collocates of the candidate domain terminology are used in the construction of a candidate ontology (see [1] for details). In my previous work I have avoided using pre-compiled dictionaries of affect and used the so-called 'local grammar' constructs for extracting patterns that were 'sentiment-laden' [3] – an approach that has allowed us to look for sentiment in texts in typologically diverse languages like English (and Urdu), Chinese and Arabic. For the purposes of comparison with other work in financial sentiment analysis, I have used the Harvard *Dictionary of Affect*: I have computed returns and volatility of affect in a corpus drawn from a representative newspaper website. My hypothesis is this: Can the computation of the volatility of affect, found in news paper reportage and editorials, help in quantifying (financial and economic) risk, much in the same manner as risk computations based on prices and values of index help in quantifying risk?

8.4.1 Corpus Preparation and Composition

The design of the corpus was motivated by the state of Irish economy during the period of 1995–2005; the first 5 years were the so-called Celtic Tiger boom eras (1996–2000) and the next 5 year period comprised the dot.com bubble, September 2001 attacks and the consequent down turn and the lead upto the introduction of the Euro. The authoritative and influential *Irish Times* that has been published since the 1850s and has a digital archive going back to 1859. One of my student (Nicholas Daly) used a text-retrieval robot to search and retrieve all items (news reports, editorials, or op-ed columns) based on the robot user: We chose the period (1995–2005) and gave the robot three keywords: *Ireland/Irish* and *Economy*. The corpus comprises 2.6 million words distributed over 4,075 news reportage and editorial items (Table 8.2):

The size of the year, viewed on an annual basis, appears to be comparable (Mean = 407, Standard Deviation = 51.522): only in 2 years both the number of stories and the verbiage was above one-standard deviation above the mean (1996 and 2001), and the number of stories in 2005 were just one s.d. above the mean (1.04).

8.4.2 Candidate Terminology and Ontology

We found that 'sentiment' in itself was a keyword and analysis of its statistically significant collocates showed that despite the boom in the late 90s the focus of *Irish Times* content was on more negative aspects, but the next 5 years show the establishment of a whole terminology nucleating around 'sentiment' (Table 8.3):

Table 8.2 Distribution of stories in our *Irish times corpus*

Year	No. of stories	No. of words	Year	No. of stories	No. of words
1996	296	1,65,937	2001	562	3,60,026
1997	395	2,59,748	2002	367	2,56,613
1998	465	2,96,531	2003	377	2,50,415
1999	447	2,95,873	2004	377	2,50,376
2000	462	3,06,063	2005	327	2,34,101
Total	2,065	13,24,152		2,010	13,51,531

Table 8.3 Compound words with 'sentiment' as a head word – a comparison over 5 year periods

1995–2000	2001–2005
▼ • sentiment	▼ • sentiment
▼ • investor_sentiment	► • business_sentiment
• factors_affecting_investor_sentiment	▼ • consumer_sentiment
▼ • market_sentiment	► • consumer_sentiment_survey
► • bond_market_sentiment	► • irish_consumer_sentiment
► • negative_sentiment	► • investor_sentiment
► • poor_sentiment	▼ • sentiment_index
	► • sentiment_index_climbed
	► • sentiment_index_produced
	► • sentiment_index_rose
	► • sentiment_index_surpassed

The above analysis was carried out using the computation of significant collocates following Frank Smadja (1993) and the assumption here was that if the word sentiment is to the left of a another word, excluding the so-called closed class words, then sentiment is the headword. The output was processed using the ontology system Protègè.

8.4.3 Historical Volatility in Our Corpus

The 2.6 million word corpus was analysed by computing the frequency of affect words in Harvard *Dictionary of Affect* (H-DoA) that were present in the texts in the corpus. The frequency was normalised for the length of the individual texts. The H-DoA comprises a large number of categories as mentioned above: we have used only two categories, *Positive* and *Negative* affect word categories that respectively have 1,916 and 2,292 words. For each news item on a given day, the frequency of all words that were labelled *Positive* and *Negative* in the H-DoA was computed. The frequency counts were aggregated on a monthly basis and returns computed. The standard deviation of the returns on annual basis was calculated and we then had *volatility* of 'positive' sentiments and that of the 'negative' sentiments.

The first thing to notice about our results is that the 'return' (change in frequency) shows much greater fluctuation in value than the frequency itself; this confirms the findings in econometrics in the context of prices and the change in prices (see, for example, 2003). This is true of both the negative and positive word frequency time-series, despite the preponderance of positive words over the negatives (Figs. 8.2 and 8.3).

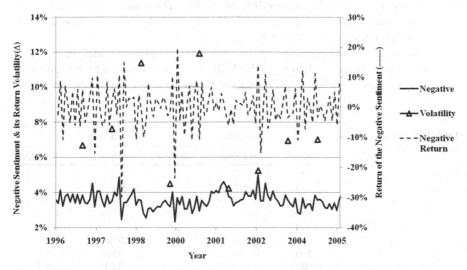

Fig. 8.2 Changes in the frequency (*full line*) of negative affect terms in our *Irish Times Corpus* (displayed monthly for 1996–2006). The returns are shown in *dashed line* (and values on the vertical axis on the right hand. The historical volatility is indicated by *solid triangles* and values are on the left-hand vertical axis

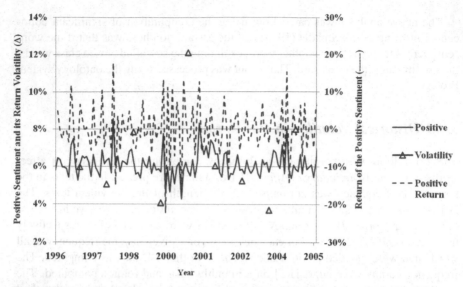

Fig. 8.3 Changes in the frequency (*full line*) of positive affect terms in our *Irish Times Corpus* (displayed monthly for 1996–2006). The returns are shown in *dashed line* (and values on the vertical axis on the right hand. The historical volatility is indicated by *solid triangles* and values are on the left-hand vertical axis

The changes in historical volatility computed over the 10 year period shows interesting results: in 1998 and 2000 the negative series had a higher than 'normal' volatility (one standard deviation above the norm) and in the two intervening years the volatility was below the norm (1999 and 2001). The positive affect series has below the norm volatility in 1999 and 2003 and much higher volatility in 2000 (2 standard deviations) (see Table 8.4).

Finally, we show the variation in the Irish Stock Exchange Index of 100 top companies listed on the Exchange (ISEQ 100). We have had access to the values of the Index on a daily basis and we have used the value at the end of the month of

		Volatility		Volatility	
	Year	Negative	Std. dev.	Positive	Std. dev.
	1996	0.064		0.057	
	1997	0.075		0.050	
	1998	0.112	1.7	0.078	
	1999	0.034	−1.2	0.038	−1.1
	2000	0.111	1.6	0.113	2.2
	2001	0.036	−1.1	0.060	
	2002	0.054		0.052	
	2003	0.054		0.037	−1.1
	2004	0.070		0.079	
	2005	0.058		0.059	

Table 8.4 Volatility changes in our two time series

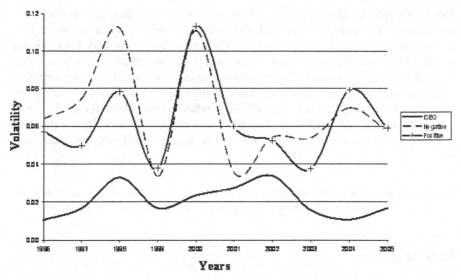

Fig. 8.4 Changes in the historical volatility in the affect series and in the ISEQ Index

each year as the ISEQ Index value and then computed returns and volatility for the period 1996–2005.

The volatility in ISEQ is smaller in comparison with that in the negative and positive affect series: this may be an artefact of computation as is the considerable variation in the volatility series of affect when compared to ISEQ (see Fig. 8.4).

8.5 Afterword

The above results give us some sense of how to find sentiment words and quantify the changes in sentiment. It is perhaps too early to read the runes: whether we can use the volatility of affect times series to compute (financial) risk. But the study looks promising. We are looking at the auto-correlation in the various time series and computing other econometric metrices to quantify changes in sentiment.

In a related study, myself and my colleagues are looking at the effect of the use of different dictionaries of affect on the measurement of sentiments, including that of the GI Dictionary. We hope to use the system for analysing reports about emerging markets and specific financial instruments (shares, derivatives, bonds) and commodities: we intend to go beyond the professional media (newspapers, company documents, stock exchange reports) and include social media (blogs, e-mails and contrarian reports). It is through the social media that the contagion affecting the stock markets spreads. This project is undertaken jointly with Trinity Business School and the Irish Stock Exchange.

A sentiment analysis based on the indirect evidence of social and professional media is only one part of the overall picture. The Trinity Sentiment Analysis Group, a multi-disciplinary group including computer scientists, linguists and economists, has launched a sentiment survey for Irish institutional and individual investors. This survey was originally developed by Robert Shiller of Yale University International Centre of Finance; we have launched this Survey in collaboration with Yale.[1] The work of the Trinity Sentiment Group is ambitious and is focussed on engendering an openness and transparency in the workings of the vitally important financial sector. We are endeavouring to bring together and synthesise inputs from the professional media, the social media, data from the stock markets, and views of the stakeholders in a common framework. This is a long term program of work which we have just begun.

Acknowledgments Dr Ann Devitt worked with me on this project and we have looked at different dictionaries of affect [8]. She had written the program that computes affect word frequency.

References

1. Ahmad, K. 2007. Artificial ontologies and real thoughts: Populating the semantic web? (Invited Keynote Speech). In *Lecture notes on artificial intelligence series (4733) – Proceedings of 10th Annual Congress of Italian Ass. of Art. Intelligence*, eds. R. Basili and M.T. Pazienza, 3–23. Berlin: Springer.
2. Ahmad, K. 2007. Being in text and text in being: Notes on representative texts. In *Incorporating Corpora: The linguist and the translator*, eds. G. Anderman and M. Rogers, 60–94. Clevedon, England: Multilingual Matters.
3. Almas, Y., and K. Ahmad. 2007. A note on extracting 'sentiments' in financial news in English, Arabic & Urdu, The Second Workshop on Computation, al Approaches to Arabic Scriptbased Languages, Linguistic Society of America 2007 Linguistic Institute, Stanford University, Stanford, California., Linguistic Society of America, pp. 1–12.
4. Althaus, S., J. Edy, P. Phalen. 2001. Using substitutes for full-text news stories in content analysis: Which text is best? *American Journal of Political Science* 45(3):707–723.
5. Bauwens, L., W.B. Omrane, and P. Giot. 2005. News announcements, market activity and volatility in the Euro/Dollar foreign exchange market. *Journal of International Money and Finance* 24:1108–1125.
6. Chang, Y., and S.J. Taylor. 2003. Information arrivals and intraday exchange rate volatility. *Journal of International Financial Markets, Institutions and Money* 13:85–112.
7. DeGennaro, R., and R. Shrieves. 1997. Public information releases, private information arrival and volatility in the foreign exchange market. *Journal of Empirical Finance* 4:295–315.
8. Devitt, A., and K. Ahmad. 2007. Cohesion-based sentiment polarity identification in financial news. The 45th annual meeting of the Association for Computational Linguistics (ACL 2007), Prague, Czech Republic, June 23–30, 2007, Stroudsburg, PA: Association of Computational Linguistics, 984–991.
9. Engle III, R. 2003. *Econometric models and financial practice.* http://nobelprize.org/nobel_prizes/economics/laureates/2003/engle-lecture.html http://en.wikipedia.org/wiki/Credit_crunch. Accessed 17th April 2008.
10. Kelly, E., and P. Stone.1975. *Computer recognition of English word senses.* Amsterdam: North-Holland Linguistic Series.

[1]https://www.cs.tcd.ie/Khurshid.Ahmad/SurveySite/

11. Kennedy, A., and Inkpen, D. 2006. Sentiment classification using contextual valence shifters. *Computational Intelligence* 22(2):117–125.
12. Lasswell, H.D. 1948. Power and personality. London: Chapman and Hall.
13. Lidén, E.R. 2006. Stock recommendations in swedish printed media: Leading or misleading? *The European Journal of Finance* 12(8):731–748.
14. Mandelbrot, B. 1963. The variation of certain speculative prices. *Journal of Business* 36: 394–419.
15. Namenwirth, Z., and H.D. Lasswell. 1970. *The changing language of American values: A computer study of selected party platforms*. Beverly Hills CA: Sage Publications.
16. Stone, P.J., D.C. Dunphy, M.S. Smith, and D.M. Ogilvie, et al. 1966. *The General inquirer: A computer approach to content analysis*. Boston:The MIT Press. (For downloading the GI Lexicon, please go to http://www.wjh.harvard.edu/~inquirer/. Accessed 17 April 2008.
17. Taylor, S.J. 2005. *Asset price dynamics, volatility, and prediction*. Princeton and Oxford: Princeton University Press.
18. Tetlock, P.C. 2007. Giving content to investor sentiment: The role of media in the stock market. *Journal of Finance* 62(3):1139–1168.

Chapter 9
Sentiment Analysis Using Automatically Labelled Financial News Items

Michel Généreux, Thierry Poibeau, and Moshe Koppel

9.1 Introduction

The aim of this research is to build on work by [7] and [15] to investigate the subjective use of language in financial news items about companies traded publicly and validate an automated labelling method. More precisely, we are interested in the short-term impact of financial news items on the stock price of companies. This is a challenging task because although investors, to a certain extent, make their decision on the basis of factual information such as income statement, cash-flow statements or balance sheet analysis, there is an important part of their decision which is based on a subjective evaluation of events surrounding the activities of a company. Traditional Natural Language Processing (NLP) has so far been concerned with the objective use of language. However, the subjective aspect of human language, i.e. sentiment that cannot be directly inferred from a document's propositional content, has recently emerged as a new useful and insightful area of research in NLP [3, 8, 16]. According to [17], affective states include opinions, beliefs, thoughts, feelings, goal, sentiments, speculations, praise, criticism and judgements, to which we may add attitude (emotion, warning, stance, uncertainty, condition, cognition, intention and evaluation); they are at the core of subjectivity in human language. We treat short financial news items about companies as if they were carrying implicit sentiment about future market direction made explicit by the vocabulary employed and investigate how this *sentimental* vocabulary can be automatically extracted from texts and used for classification. There are several reasons why we would want to do this, the most important being the potential of financial gain based on the exploitation of covert sentiment in the news items for short-term investment. On a less pragmatic level, going beyond literal meaning in NLP would be of great theoretical interest for language practitioners in general, but most importantly perhaps, it would be of even greater interest for anyone who wishes to get a sense of what are people feelings

M. Généreux (✉)
Laboratoire d'informatique de Paris-Nord, Université Paris 13, 93430 Villetaneuse, France;
Departamento de Linguística Geral e Românica, Faculdade de Letras da Universidade, de Lisboa
Alameda da Universidade, 1600-214 Lisboa, Portugal
e-mail: genereux@clul.ul.pt

K. Ahmad (ed.), *Affective Computing and Sentiment Analysis*, Text, Speech
and Language Technology 45, DOI 10.1007/978-94-007-1757-2_9,
© Springer Science+Business Media B.V. 2011

towards a particular news item, topic or concept. To achieve this we must overcome problems of ambiguity and context-dependency. Sentiment classification is often ambiguous (compare *I had an accident* (negative) with *I met him by accident* (not negative)) and context dependent (*There was a decline*, negative for *finance* but positive for *crimes*).

9.2 Data and Method

The automated labelling process is described in Section 9.2.4. We have opted for a linear *Support Vector Machine (SVM)* [4] approach as our classification algorithm and we have used the Weka[1] software package.

9.2.1 Training and Testing Corpus

Based on previous work in sentiment analysis for domains such as movie reviews and blog posts, we have selected an appropriate set of three key parameters in text classification: feature *type, threshold* and *count*. Our goal is to see whether the most suitable combinations usually employed for other domains can be successfully transferred to the financial domain. Our corpus is a subset of the one used in [7]: 6,277 news items averaging 71 words covering 464 stocks listed in the Standard & Poor 500 for the years 2000–2002.

9.2.2 Feature Types

We consider five types of features: *unigrams, stems, financial terms, health-metaphors* and *agent-metaphors*. The news items are tokenized with the help of a POS tagger [14]. *Unigrams* consist of all nouns, verbs, adjectives and adverbs[2] that appear at least three times in the corpus. *Stems* are the unigrams which have been stripped of their morphological variants. The *financial terms* stem from a 'clinical study of investors discussion and sentiment' [2]. The list comprises 420 words and their variants created by graduate students who read through messages[3] and selected words they felt were relevant for finance (not necessarily most frequent).[4] *Health metaphors* are a list of words identified by [6] in a 6 million word corpus from the *Financial Times* suggesting that the financial domain is pervaded by terms from the medical domain to describe market phenomena: examples include *addiction, chronic* and *recovery*. The full list comprises 123 such terms.

---

[1] http://www.cs.waikato.ac.nz/ml/weka/
[2] This list is augmented by the words *up, down, above* and *below* to follow [7].
[3] The corpus was a random selection of texts from Yahoo, Motley Fool and other financial sites.
[4] Sanjiv Das, *personal communication*.

Finally, recent work by [9] suggests that in the case of market trends, investors tend to process *agent metaphors*, when the language treats the market as though it were an entity that produces an effect deliberately (e.g. *the NASDAQ climbed higher*), differently from object metaphors, where the language describe price movements as object trajectories, as events in which inanimate objects are buffeted by external physical forces (e.g. *the Dow fell through a resistance level*) or non-metaphorical expressions that describe price change as increase/decrease or as closing up/down (e.g. *the Dow today ended down almost 165 points*). The same study gives the verbs *jump, climb, recover* and *rally* as the most frequent indicators of uptrend movement, and *fall, tumbled, slip* and *struggle* as the most frequent indicators of downtrend movements. The point made in the study is that in the case of agent metaphors, investors tend to believe that the market will continue moving in the same direction, which is not the case for object metaphors or non metaphors. These results are potentially useful for sentiment analysis, as we are trying to find positively correlated textual features with market trends. To construct a list of potential agents, we extracted all nouns from our corpus and used WordNet[5] to filter out elements which were not hyponym of the synset comprising causality and agency words,[6] defined as *an entity that produces an effect or is responsible for events or results*: in this way we collected 553 potential agents. To allow those agents to carry out their actions, we completed this list with all 1,538 verbs from the corpus.

9.2.3 Feature Selection and Counting Methods

We consider three feature selection methods that [18] reported as providing excellent performance. *Document Frequency* (DF) is the number of documents in which a term occurs. We computed DF for each feature and eliminated features for which DF fell below a threshold (100). In *Information Gain* (IG), features are ranked according to a preferred sequence allowing the classifier to rapidly narrow down the set of classes to one single class. We computed the 100 features with the highest information gain. Finally, the χ^2 statistic measures the lack of independence between a feature and a set of classes. We computed the top 100 least independent features. It is worth mentioning that the same 100 features were selected using either IG or χ^2 statistic, except for a few features ranking order in the top ten.

There are different methods for *counting* values of the features mentioned above. There are two methods worth considering for computing the value of a given feature when a token with the feature is found: the first is the binary method where a value of zero indicates the absence of the feature whereas a value of one indicates the presence of the feature. This method appears to yield good results for movie reviews [12]. The second simply gives a count of the feature in the document and normalises the count for a fixed-length document of 1,000 words (TF).

---

[5] http://wordnet.princeton.edu/
[6] Wordnet synset number 100005598: *causal agency#n#1, cause#n#4* and *causal agent#n#1*

9.2.4 News Items and Stock Price Correlation

To construct our 500 positive examples we used similar criteria as [7], based on contemporaneous price changes: the price of a stock was noted at the opening of a market after a news item was published; and, the price of the stock was noted at the closing of the market on the day before a news item was published. Essentially, for a news item to be labelled as a positive example, its positive price change must be greater than a given threshold (we used 4%) and be in excess of the overall S&P index change.

For instance, the following news item about the company *Biogen, Inc.* (symbol BGEN), appeared on May 23rd 2002:

Biogen, Inc. announced that the FDA's Dermatologic & Ophthalmic Drug Advisory Committee voted to recommend approval of AMEVIVE (alefacept) for the treatment of moderate-to-severe chronic plaque psoriasis.

At the opening on the 24-May-2002, the price reached $48.43, whereas at the closing on the 22-May-2002 the price was $38.71. Therefore, there is a positive price change of

$$\frac{\$48.43 - \$38.71}{\$38.71} = 0.19995$$

or almost 20%, so the news item is classified as being positive. The same reasoning is applied to find 500 negative examples, corresponding to a negative price change of at least 4%.

The corpus of positive and negative news items, contemporaneous with price changes, was used in the training. We have five feature types, three feature selection methods, and two counting methods – 30 different ways to represent news items. To narrow down the training possibilities, we first focused on the features themselves. The *unigram* feature, in combination with the *information gain* feature selection criterion and *binary count* of the feature values, appear to give the highest accuracy (67.5%) when compared to the use of other features including *stems* and the *agent-metaphors, financial terms* and *health-metaphors* respectively (see Table 9.1). The method of evaluation was 10-fold cross-validation on the news items dataset.

The unigram feature, with its inherent simplicity, appears to perform well in combination with certain feature selection and feature counting parameters, We found that the highest classification accuracy, at 67.6%, was obtained by using *unigrams, information gain (IG)* and *term frequency (TF)* (Table 9.2). The second highest

Feature	Accuracy (%)
Unigram	67.5
Stems	66.9
Agent metaphor	66.4
Financial metaphor	59.2
Health metaphor	52.4

Table 9.1 Feature tuning: Performance of various features with information gain and binary count

Table 9.2 Feature tuning: Performance of unigram features with different feature selection and counting methods

Feature	Feature selection	Feature count	Accuracy (%)
Unigram	Information gain	Term frequency	67.6
Unigram	Information gain	Binary count	67.5
Unigram	χ^2	Binary count	66.1
Unigram	Degrees of freedom	Binary count	59.4

accuracy was obtained by using *unigrams, information gain (IG)* and *binary count (Bin)* and reached 67.5%. Given this small, perhaps insignificant difference in accuracy together with the favourable reviews of the binary method in the literature, we believe that this basic trio (*unigrams, IG and binary*) will suffice.

Our results suggest that the features based on a list of agent metaphors describing market trend movements appear more useful for the classification of financial news items than a list of health metaphors or a human-constructed list of financial terms. At closer examination, it appears that most of the contribution is made by the notion of *agent*: only five of the eight most frequent indicators (*recover, climb, fall, slip* and *struggle*) actually appear in our corpus, and only one (*fall*) made the cut through the top 100 features that bring most information gain. We conjecture that the description of financial news items retains the same agent-based feature as in market trend description, however it is expressed by commentators using a different set of (predicative) terms. In the remaining experiments we depart slightly from [7] by taking into account negation, i.e. negated words (e.g. not rich) are featured as a single term (not_rich). We also excluded proper nouns as a potential feature: In financial news items, proper nouns usually refer to companny names and employees at a specific instance of time and these nouns change over time as new companies and people enter the domain and many leave. Proper nouns do not appear to us as an appropriate feature.

9.2.5 Feature Selection and Semantic Relatedness of Documents

A study by [10] has suggested that information from different sources can be used advantageously to support more traditional features. Typically, these features characterise the semantic orientation (SO) of a document as a whole [5, 11]. One such feature is the result of summing up the semantic relatedness (Rel) between all individual words (adjectives, verbs, nouns and adverbs) with a set of polarised positive (P) and negative (N) terms, for the domain of interest, here finance. The semantic relatedness of a document can be defined as:

$$\sum_{w}^{Words} \left(\sum_{p}^{P} Rel(w, p) - \sum_{n}^{N} Rel(w, n) \right)$$

Table 9.3 Polarised terms used in our study

Adjectives		Nouns		Verbs		Adverbs	
Positive	Negative	Positive	Negative	Positive	Negative	Positive	Negative
Good, rich	Bad, poor	Goodness, richness	Badness, poverty	Increase, enrich	Decrease, impoverish	Well, more	Badly, less

Note that the quantity of positive terms P must be equal to the quantity of negative terms N. To compute relatedness, we used the method described in [1] and Word-Net.[7] The list of polarised terms we used are presented in Table 9.3.

Our experiments suggest that the relatedness measure produces a negative semantic orientation even for documents that are labelled as positive. Even more worryingly, the negative score of positively labelled documents is greater than the negative score of a negatively labelled documents. This suggests that the polarity orientation of financial news items cannot be captured by techniques based on semantic relatedness and our result shows no significant improvement on accuracy (69%) if we include semantic relatedness as one of our features.

9.3 Results

9.3.1 Horizon Effect

The next experiment looks at the lasting effect of a news item on the stock price of a company. Using 300 positive examples and 300 negative examples with a ±2% price variation, we computed classification accuracies for non-cotemporeneous, more precisely subsequent, price changes. Therefore, news items were classified according to price changes from the opening the first open market day after the news to X number of days after the news. We consider the following values for X: 2, 3, 7, 14 and 28. (See Table 9.4). Given that classification accuracies are slowly worsening as we move further away from the day the news item first broke out (coefficient of correlation is −0.89), we conclude that some prices are getting back to, or even at the opposite of, their initial level (i.e. before the news broke out). Assuming that in the interval no other news items interfered with the stock price, this result also reinforced the validity of the automatic labelling technique.

Table 9.4 Horizon effect

Horizon	Accuracy (%)
[+1,+2]	69.5
[+1,+3]	68.8
[+1,+7]	67.5
[+1,+14]	68.0
[+1,+28]	66.3

---

[7] Using the PERL package [13].

9.3.2 Polarity Effect

This experiment looks at the effect on accuracy a change in the labelling distance between two classes produces. The intuition is that the more distant two classes are from each other, the easiest it is for the classifier to distinguish among them, which translates as a higher accuracy.

To test the accuracy of our classifier, we created five sub-corpora each comprising 400 labelled news items: We have three sub-corpora which exclusively contain positive (price change greater than $+2\%$), negative (price change smaller than -2%) or neutral text respectively, the other two corpora are a mixture of negative and neutral news items and positive and neutral news items respectively (Table 9.5).

One can imagine a line of unit length along which we can classify the subcorpora in Table 9.5: The sub-corpus I, with purely negative texts, lies at the origin and sub-corpus V, with purely positive news items, lies at unit distance from the subcorpus I. The *distance* between corpora I and II (and II and III, III and IV, IV and V) is assumed to be 0.25; the distance between I and III (neutral news items) is 0.5 and I and IV (positive plus neutral news items) is 0.75 and so on. When our classifier is presented with a mixture of subcorpora with greatest distance, then its accuracy should be the best and contrariwise for a mixture of subcorpora that lie close to each other will lead to reduced accuracy of classification. The *distance* between two subcorpora changes the classification accuracy by 8% on average: when the mixture is of proximate subcorpora (*distance* of 0.25) the accuracy is over 62% and rises to 70% when the distance approaches unity (Table 9.6). This is further evidence of the utility and accuracy of the automatic labelling technique.

Table 9.5 Subcorpora for testing polarity effects

Sub-corpus	Negative	Neutral	Positive	Total
I Negative only	400			400
II Neg+Neu	200	200		400
III Neutral only		400		400
IV Neu+Pos		200	200	400
V Positive only			400	400

Table 9.6 Accuracy for mixture classification

Class 1	Class 2	Accuracy	Nominal distance
Neg+Neu	Pos	0.703	0.75
Neg	Pos	0.698	1
Neu	Pos	0.693	0.5
Neg	Neu+Pos	0.693	0.75
Neg	Neu	0.68	0.5
Neg+Neu	Neu	0.646	0.25
Neu+Pos	Pos	0.641	0.25
Neg	Neg+Neu	0.628	0.25
Neg+Neu	Neu+Pos	0.618	0.5
Neu	Neu+Pos	0.576	0.25

9.3.3 Range Effect

The range effect experiment explores how the size of the minimum price change for a news item to be labelled either as positive or negative influences classification accuracy. The intuition is that the more positive and negative news items are labelled according to a larger price change, the more accurate classification should be. Table 9.7 shows results using cotemporeneous price changes. The labelling method once again yields expected results: for two classes (positive and negative), the more comfortable the price change margin gets, the more accurate classification is (coefficient of correlation is +0.86). However, accuracies appear to reach a plateau at around 67%, and classification accuracy improvements beyond 75% seems out of reach. The last column of Table 9.7 reports accuracies for the case where news items whose price change is falling between the range are labelled as *neutral*. Although accuracies are, as expected, lower than for two classes, they are significantly above chance (33%). The same positive correlation is also observed between the price change margin and accuracies (coefficient of correlation is +0.88). In the next experiment we examine more in depth the effect of adding a neutral class on precision.

9.3.4 Effect of Adding a Neutral Class on Non-cotemporaneous Prices: One- and Two-Days Ahead

In all but one of the experiments so far, we have considered classes with maximum polarity, i.e. with a neutral class separating them. On the one hand this has simplified the task of the classifier since news items to be categorised belonged to one of the positive or negative extremes. On the other hand, this state of affairs is somewhat remote from situations occurring in real life, when the impact of news items can be limited. Moreover, the information about overall accuracy of classification is not the most sought after information by investors. Let's examine briefly more useful information for investors:

> *Positive Precision* A news item which is correctly recognised as positive is a very important source of information for the investor. The potential winning strategy now available is to buy or hold the stock for the corresponding range. Therefore, it is very important to build a classifier with high precision for

Table 9.7 Range effect

Range	Nb examples	2-class (%)	3-class (%)
±0.02	1000	67.8	46.3
±0.03	1000	67.1	47.9
±0.05	800	69.5	46.8
±0.06	600	74.0	50.1
±0.07	400	76.3	50.1
±0.10	200	75.0	51.3

the positive class, significantly above 50% to cover comfortably transaction
costs.

Negative Precision A news item which is correctly recognised as negative is
  also an important source of information for the investor. The potential saving
  strategy now available to the investor, given that he or she owns the stock,
  is to sell the stock before it depreciates. Therefore, it is important to build a
  classifier with high precision for the negative class, significantly above 50%
  to cover safely transaction costs.

Positive and Negative Recall Ideally, all positive and negative news items
  should be recognised, but given the potential substantial losses that misrecog-
  nition (implying low positive/negative precision) would imply for investors,
  high recall should not be a priority.

Table 9.8 gives an indication of the kind of effect positive (+precision) and negative (−precision) precision we can expect if we built a 3-class classifier. Results
show that precision is either worryingly close to 50% (the positive case), or is very
volatile and could swing the precision level well below 50% on too many occasions.
This perhaps demonstrates that if we were to build a financial news items classifier
satisfying at least high precision for the positive news items, it appears important to
avoid the three-class classification approach.

9.3.5 Conflating Two Classes

In Section 9.3.4 we underlined the importance of high precision for the classification of positive and negative news items and concluded that a 3-class categoriser
was unlikely to satisfy this requirement. In this section we conflate two of the three
classes into one and examine the effect on precision and recall. Table 9.9 displays
three classification measures for the case where the classes neutral and negative
have been conflated to a single class. Table 9.10 displays three classification measures for the case where the classes neutral and positive have been conflated to
a single class. We used a range of ±0.02, a forward-looking horizon of [+1,+2]
days with 800 training examples. It is difficult to evaluate precisely what the cost of

Table 9.8 Effect of adding a neutral class on non-cotemporaneous prices

Range	Nb examples	−Precision (%)	+Precision (%)
±0.01	1000	77	51
±0.02	800	41	53
±0.03	400	69	54

Table 9.9 Positive versus combined neutral and negative news items

Measure/Class	POS	NEG+NEU
Precision	0.857	0.671
Recall	0.555	0.908
Accuracy	0.7313	

Table 9.10 Negative versus combined neutral and positive news items

Measure/Class	NEG	POS+NEU
Precision	0.652	0.805
Recall	0.870	0.535
Accuracy	0.7025	

trading represents, but there seems to be enough margin of maneuver to overcome this impediment, especially in the case of the positive classifier (Table 9.9).

9.3.6 Positive and Negative Features

Closer examination of the features resulting from the selection process paints a different picture from the one presented Koppel and Shtrimberg [7]. These authors have used all words that appeared at least sixty times in the corpus, eliminating function words with the exception of some relevant words. We kept only adjectives, common nouns, verbs, adverbs and four relevant words, *above, below, up* and *down*, that appear at least three times in the training corpus. In a nutshell, [7] found that there were no markers for positive stories, which were characterised by the absence of negative markers. As a result, recall for positive stories were high but precision much lower. Our findings are that negative and positive features are approximately equally distributed (53 negatives and 47 positives) among the top 100 features with the highest information gain and that recall and precision for positive stories were respectively lower and higher. We define a *polarity* orientation (positive or negative) of each feature as the class in which the feature appears the most often. Table 9.11 shows the top ten positive features and Table 9.12 the top ten negative features. The *Rank* column indicates the position of the feature in the top 100 ranking resulting from the information gain screening. The $+df/-df$ column displays the number of documents (examples) in which the feature appears at least once ($+df$ for positive and $-df$ for negative). The $+tf/-tf$ column displays the number of times the feature appears in the entire set of documents ($+tf$ for positive and $-tf$ for negative), while the $+n/-n$ column displays the same values normalised to a constant document length of 1,000 words. For example, the feature *common* appears in 29 positive

Table 9.11 Positive features

Rank	Feature	+df/−df	+tf/−tf	+n/−n
1	Common	29/8	33/13	1318/390
2	Shares	33/11	48/17	2014/640
3	Cited	20/4	20/4	427/49
5	Reason	18/4	18/4	411/69
8	Direct	7/0	7/0	163/0
9	Repurchase	15/3	26/3	818/115
10	Authorised	17/4	18/5	596/177
11	Drug	6/0	6/0	114/0
13	Partially	6/0	6/0	89/0
14	Uncertainty	6/0	6/0	102/0

Table 9.12 Negative features

Rank	Feature	+df/−df	+tf/−tf	+n/−n
4	Change	0/8	0/11	0/206
6	Work	1/11	1/12	33/183
7	Needs	0/7	0/7	0/139
12	Material	0/6	0/6	0/128
15	Pending	0/6	0/7	0/100
16	Gas	8/23	13/34	229/571
19	Cut	1/9	1/10	15/216
20	Ongoing	1/9	2/14	14/201
25	E-mail	0/5	0/5	0/68
26	Week	0/5	0/5	0/72

examples and 8 negative examples. It also appears 33 times in all positive examples and 13 times in all negative examples. Below is one highly positive news item (+11% price change) and one highly negative news item (−49% price change) with positive features inside square brackets and negative features inside braces. The following news item about the company Equifax Inc. (symbol EFX) appeared on the 20th of September 2001. Its stock price jumped from $18.60 at opening on the 21st of September 2001 to $20.70 on the 24th of September 2001, for a price change of 11.29%:

> Equifax Inc. announced that it is repurchasing [shares] in the open market, pursuant to a previous [repurchase] authorisation. The [Company]'s board of directors had [authorised] a repurchase of up to $250 million of [common] stock in the open market in January 1999, of which approximately $94 million remains available for purchase.

The following news item about the company Applied Materials, Inc. (symbol AMAT) appeared on the 15th of April 2002; its stock price plummeted from $53.59 at opening on the 16th of April 2002 to $27.47 on the 17th of April 2002, for a price change of −48.74%:

> Applied Materials, Inc. announced two newly granted U.S. Patents No. 6,326,307 and No. 6,362,109, the [Company]'s third and fourth patents covering the use of hexafluorobutadiene (C4F6) {gas} chemistry for critical dielectric etch applications. A high-performance etch process chemistry, C4F6 used in an Applied Materials etch system, enables the industry's move to the 100nm chip generation and beyond.

9.4 Discussion

The surprisingly encouraging results we have presented for a forward-looking investment strategy should not be viewed outside its specific experimental setup conditions. In what follows we highlight a number of points worth considering:

9.4.1 Lack of Independent Testing Corpus

Cross-validation is a method which can provide a solid evaluation of the overall accuracy of a classifying method. However, a more accurate evaluation should involve an independent testing corpus, ideally covering a distant time-period to avoid overfitting or over-training. Nevertheless, we have attempted to avoid these caveats by keeping a small number of features compared to the number of training examples and by avoiding the use of proper nouns as features.

9.4.2 Pool of Features

Our pool of features was selected among the entire training set, which includes the cross-validated sections. Although to a small degree, this may have caused a *data-snooping* bias, where features were selected among the testing examples. On the other hand, as can be observed in Tables 9.11 and 9.12, the interpretation of positive and negative features is not straightforward, which suggests that portability among different domains and even time periods could be problematic.

9.4.3 Size of Documents

Clearly, the size of documents is crucial for classification. The corpus we used averaged just over 71 words per document, which in general should be long enough to collect enough statistics. Nevertheless, if we look at our top ten positive stories (those with the highest positive price change), we found that half of them contained no feature at all, whereas three out of our top ten negative examples were similarly deprived of features. Given that this situation is likely to worsen if we train and test on different domains and periods, this is a potential area where a default bias can be difficult to avoid (i.e. a document without features will systematically be classified in the same class). One solution would be to increase the number of features.

9.4.4 Trading Costs

If the minimum transaction level to overcome fixed and relative trading costs is high, this brings upon the investors a burden of risk which he or she may not be able or willing to bear. The classifier should be characterised clearly by its level of precision matched with an estimate of the trading costs that would guide the investor in its decision.

9.5 Conclusion and Future Work

We have revisited a method for classifying financial news items using automatically labelled data. Our findings give a different picture of the set of features best suited for the task and a somewhat less pessimistic prognosis as to the validity of such an approach for forward-looking investment. We have suggested some topics where further research should be carried out for testing the automatic labelling approach within a practical and realistic framework. To this end, our next step will be to use our system coupled with a virtual trading site[8] to monitor financial news items with a view to use the analysis for investing in companies. This should give us a better idea of the effect of the transaction costs as well as the portability of the features and model developed during our experiments.

References

1. Banerjee, S. and T. Pedersen. 2003. Extended gloss overlaps as a measure of semantic relatedness. In Proceedings of the Eighteenth International Joint Conference on Artificial Intelligence, 805–810, Acapulco.
2. Das, Sanjiv, Asis Martinez-Jerez, and Peter Tufano. 2005. E-information: A clinical study of investor discussion and sentiment. *Financial Management* 34(5):103–137.
3. Ann Devitt, and Khurshid Ahmad. 2007. Sentiment polarity identification in financial news: A cohesion-based approach. In Proceedings of ACL-07, the 45th Annual Meeting of the Association of Computational Linguistics, 984–991, Prague, CZ, June 2007. ACL.
4. Joachims, Thorsten. 2001. *Learning to classify text using support vector machines*. Boston: Kluwer Academic Publishers.
5. Kamps, J., M. Marx, R. Mokken, and M. de Rijke. Using Wordnet to measure semantic orientation of adjectives. *In LREC 2004* IV:1115–1118.
6. Knowles, Francis. 2004. Lexicographical aspects of health metaphors in financial texts. In Proceedings Part II of Euralex 1996, pp. 789–796, Department of Swedish, Göteborg University, 1996.
7. Koppel, Moshe, and Itai Shtrimberg. Good news or bad news? Let the market decide. In AAI Spring Symposium on Exploring Attitude and Affect in Text, pp. 86–88. Stanford University, March 2004.
8. Mishne, Gilad. 2007. Applied text analytics for blogs. PhD thesis, University of Amsterdam.
9. Morris, Michael W., Oliver J. Sheldon, Daniel R. Ames, and Maia J. Young. 2007. Metaphors and the market: Consequences and preconditions of agent and object metaphors in stock market commentary. *Journal of Organizational Behavior and Human Decision Processes* 102(2):174–192, March 2007.
10. Mullen, Tony, and Nigel Collier. 2004. Sentiment analysis using support vector machines with diverse information sources. In Empirical Methods in NLP.
11. Osgood, Charles E., George J. Suci, and Percy H. Tannenbaum. 1957. The Measurement of Meaning. University of Illinois.
12. Pang, Bo, Lillian Lee, and Shivakumar Vaithyanathan. 2002. Thumbs up? Sentiment classification using machine learning techniques. In Proceedings of the 2002 Conf. on Empirical Methods in Natural Language Processing.
13. Pedersen, Ted. 2004. Wordnet::similarity – measuring the relatedness of concepts. In *Proceedings of the Nineteenth National Conference on Artificial Intelligence (AAAI-04)*.
14. Schmid, Helmut. 1994. Probabilistic part-of-speech tagging using decision trees. In Int. Conference on New Methods in Language Processing, Manchester.
15. Lawrie, D., P. Ogilvie, D. Jensen, V. Lavrenko, M. Schmill, and J. Allan. 2000. Mining of concurrent text and time series. In 6th ACM SIGKDD Int. Conf. on Knowledge Discovery and Data Mining, August 2000.
16. Shanahan, J.G., Y. Qu, J. Weibe, eds. 2006. *Computing attitude and affect in text: Theory and applications*. Dordrecht: Springer.

---

[8] http://vse.marketwatch.com/

17. Wilson, T., and J. Wiebe. 2003. Annotating opinions in the world press. In *Proceedings of SIGdial-03*, 13–22.
18. Yang, Yiming, and Jan O. Pedersen. A comparative study on feature selection in text categorization. In Douglas H. Fisher, editor, Proc. of ICML-97, 14th Int. Conf. on Machine Learning, pp. 412–420, Nashville, US, 1997. Morgan Kaufmann Publishers, San Francisco.

Chapter 10
Co-Word Analysis for Assessing Consumer Associations: A Case Study in Market Research

Thorsten Teichert, Gerhard Heyer, Katja Schöntag, and Patrick Mairif

10.1 Introduction

With the advancement of natural language processing (NLP), many new research opportunities have opened up for scientists in various different fields. Among these, marketing research constitutes a prominent but yet unexplored application field. Sentiment analysis is particularly relevant in marketing contexts because it is essential for deriving an in-depth understanding of consumer behaviour. Recent innovative approaches in this area rely upon in-depth interviewing to gain insight into consumers' thoughts and feelings regarding specific brands and products [12]. Interviews yield a large amount of qualitative data that is hard to handle and needs to be structured in order to be analyzed. Manual coding and categorization can be cumbersome and time consuming. Therefore, the development and application of text analysis software is of high importance for marketing researchers both on the academic and the practitioners' side.

This manuscript illustrates an exemplary best-practice case study for the application of text analysis tools. The presented case analyzes the associations of female consumers with the product category "shoes". This product category is assumed to be emotionally laden especially for female consumers, reaching far beyond mere functional aspects. Data elicitation and processing techniques are based on methods derived from Human Associative Memory models and network analysis. As opposed to many text analysis applications, data are not obtained from secondary (internet) sources but from 30 personal in-depth interviews with female consumers. The underlying objective for the pursued marketing research is to derive a novel characterization of female shoe consumers. This should build the basis for developing innovative marketing measures which target yet unexplored consumer sentiments.

Automated text analysis is used to identify features and structures from the qualitative data at hand. Text analysis tools are integrated into qualitative data analysis

T. Teichert (✉)
University of Hamburg, Marketing and Innovation, D-20146 Hamburg, Germany
e-mail: teichert@econ.uni-hamburg.de

K. Ahmad (ed.), *Affective Computing and Sentiment Analysis*, Text, Speech
and Language Technology 45, DOI 10.1007/978-94-007-1757-2_10,
© Springer Science+Business Media B.V. 2011

in order to minimize subjectivity and to maximize replicability by excluding all elements of personal experience and emotions from the coding process. The results of the automated text analysis are contrasted with manual feature coding, showing a comparable coding quality while yielding considerable savings of time and effort. Thus we conclude that NLP offers a high potential for future research applications to solve marketing problems.

10.2 Conceptual Background

The specific requirements for text analysis tools are derived from theories and techniques underlying the applied elicitation and analysis techniques. Modellers need to understand the working of consumer memory as well as the encountered problems in eliciting and analyzing qualitative data in order to develop an appropriate procedure of automated text analysis.

10.2.1 Consumer Associations and Mental Processing

Various theories and models exist that explain the working of human memory. Particularly in the field of marketing and consumer behaviour research, Human Associative Memory (HAM) is a widely accepted model with an increasing number of studies based upon it [5, 9]. According to this model, information is stored in nodes which are linked (associated) with each other forming a complex network of associations [1, 8]. Based upon this, Spreading Activation theory provides a (however not uncriticized) framework to explain temporal aspects of associations [6]. It assumes that mental activity spreads from active concepts to all related concepts.

In the case of brands, for instance, the stimulating element can be a brand's logo or advertising jingle: individual nodes within the brand's associative network are activated and become accessible and retrievable. Activation then spreads to adjacent nodes turning activated nodes into source nodes which, in turn, spread their activation to their neighbour nodes [1, 4]. This spread of activation produces a chain, or flow, of thoughts. A representation of this flow of thoughts, though inevitably incomplete, can be obtained from the flow of speech, e.g. when eliciting brand or product associations during an interview. Speech not only contains the main aspects that are stored for a particular concept, i.e. the informational content of nodes, it can also be used to track the flow of thought and thus the existing associations, i.e. links between nodes, in the interviewee's mind.

Since most information is stored non-verbally in the human mind, standardized questionnaires and straightforward questioning often do not produce the desired results. Elicitation techniques, such as the Zaltman Metaphor Elicitation Technique [13], take the non-verbal nature of human knowledge into account and aim at surfacing primary and secondary associations. Applying visual, projective, and sensory techniques helps access subconscious memory of episodic, autobiographic, visual, and sensory nature as well as a metaphoric description of thoughts, sentiments, and emotions [12].

10.2.2 *Drawbacks of Manual Data Analysis*

The scientific discourse reveals several basic problems of qualitative data analysis. When analyzing the flow of speech in the form of transcribed interviews, researchers are frequently faced with ambiguity of statements and expressions. In order to structure and code the text on hand, they are required to interpret the interviewees' statements resulting in a rather subjective representation of the elicited data. The replicability of the results is consequently rather low. An existing and widely used remedy to this problem is the parallel data processing and coding by two or more researchers. However, when there are differences between raters, codes are often assigned based on discussions which constitute another subjectivity factor.

The higher the inter-rater-reliability, i.e. the percentage of identical codes among the raters given independently of each other, the higher the objectivity level is assumed to be. By convention, inter-rater-reliabilities of 70% and above are acceptable. Previous evidence reveals context dependencies with respect to the achievable coding reliability. Particularly in the case of sentiment coding, it can be hypothesized that inter-rater-reliability is comparatively low for emotional aspects as opposed to more rational expressions.

Text analysis tools offer a solution to this problem as they reduce the level of subjectivity to a minimum, both during the feature extraction and the categorization processes. This leads to a high replicability level and, thus, to a higher level of reliability. Following this, we expect especially high value contributions of automated text analysis for assessing consumer sentiments.

10.2.3 *Requirements for Automated Co-Word Analysis*

A tool was to be developed that addressed the particular requirements of sentiment analysis based on qualitative interviews with real-life consumers. The concept of Human Associative Memory guided the data processing and evaluation process by four main assumptions:

1. Words or concepts mentioned together are linked in the mind.
2. The more salient a concept is, the more often it is mentioned during the course of an interview.
3. The stronger the association between two concepts, the more often they are mentioned together.
4. Valence of a concept is indicated by positive or negative adjectives annotated to it.

Qualitative interview data consist of lengthy, often quite unfocused text information. Thus, text analysis tools first and foremost need to identify and extract the main consumer thoughts and sentiments from the transcribed interviews while excluding irrelevant aspects. Further, in order to make the sentiment data more manageable and interpretable, mentioned sentiments need to be assigned to categories that represent shared types of feelings toward the product or brand in question. Finally, data need to be coded to apply quantitative network analysis.

In sum, the specific requirements of the text analysis tool are as follows:

1. Extraction of features and consolidation of extracted features into meaningful categories.
2. Processing of the data using a co-word-analysis (on a paragraph level) as basis for the development of associative networks.
3. Consideration of valence expressions for the weighting of individual features.

The developed text analysis tool fulfils all of the above mentioned criteria while providing an intuitive user interface for marketing researchers without extensive IT background.

10.3 Technique and Implementation

The complete process covered by the tool comprises the following steps:

1. Import of text sources.
2. Processing of text.
3. Graph creation.
4. Graph clustering.

If there is a large amount of text to be processed, all processing steps can be done by an automatic batch process. To play with the data and tune parameters, the user is provided with a graphical user interface that reflects the processing structure and allows them to interfere at each processing step (cf. Fig. 10.1).

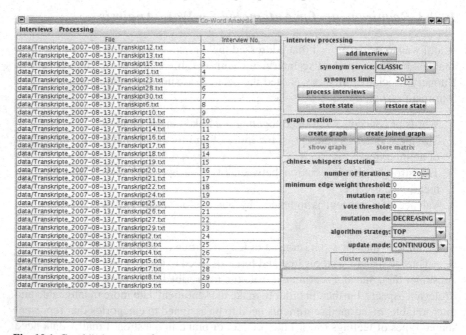

Fig. 10.1 Graphical user interface

10.3.1 Import of Text Sources

The first task covers the reading of text files with annotated metadata in a specific file format. Texts are automatically divided into sections and paragraphs that contain answers to questions as part of interviews.

10.3.2 Processing of Text

The second task does the real work: the processing of text. The aim is to extract features for sections and paragraphs, to reduce word forms to base forms and to add synonyms to each of them.

The extraction of features is mainly pattern based. First, stop words are eliminated and valence words are recognized. Features are then supposed to be located right of valence words. Stop words are the usual high frequency general language and domain specific uninteresting words with the exception of valence markers. These are words that modify the meaning of the features. E.g. in a sentence such as "These shoes are very comfortable", the words "these" and "are" are recognized as stop words, "very" is a valence word and "shoes" and "comfortable" are features; "very" is associated to and located left of "comfortable".

Each valence word has an associated value that modifies the value of the feature. These values can be positive or negative, e.g. "don't" would be considered a valence word with negative value. Each sentence is processed separately in order to avoid side effects with valence words at the end of a sentence. The text is searched for known valence words first (to ensure that we do not eliminate valence words if they appear in the list of stop words). Next, stop words are eliminated. What remains are the features and possibly valence words associated to them.

The resulting features are reduced to their base forms by using a web service for base form reduction offered by the department for natural language processing at the University of Leipzig (http://wortschatz.uni-leipzig.de/Webservices/). To get the base form of a given word, the web service simply uses a pre-processed list of words that associates with each word form a corresponding base form [2].

The base forms of the features are then used to request synonyms from the Leipzig web services. As a result, a list of weighted features is derived with a list of synonyms associated to each of them.

To make the data visible, HTML documents are generated for each original text, highlighting the extracted information (cf. Fig. 10.2). The features are marked in green or red with colour shades from dark to light green for positive and red for negative values; valence words are printed in italics; the base forms are inserted in square brackets behind the feature and synonyms are shown as tool tips.

10.3.3 Graph Creation and Clustering

In order to cluster the features, we build a graph that associates similar features. To this end, the synonym vectors of the features are compared and a similarity value is

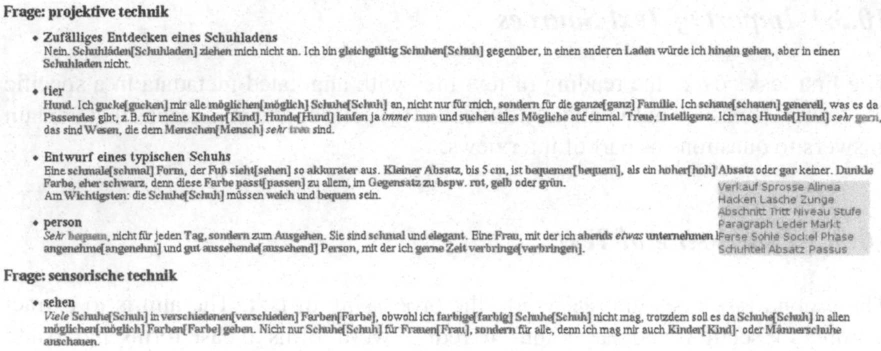

Frage: projektive technik

- Zufälliges Entdecken eines Schuhladens
  Nein. Schuhläden[Schuhladen] ziehen mich nicht an. Ich bin gleichgültig Schuhen[Schuh] gegenüber, in einen anderen Laden würde ich hinein gehen, aber in einen Schuhladen nicht.

- tier
  Hund. Ich gucke[gucken] mir alle möglichen[möglich] Schuhe[Schuh] an, nicht nur für mich, sondern für die ganze[ganz] Familie. Ich schaue[schauen] generell, was es da Passendes gibt, z. B. für meine Kinder[Kind]. Hunde[Hund] laufen ja *immer* *nun* und suchen alles Mögliche auf einmal. Treue, Intelligenz. Ich mag Hunde[Hund] *sehr gern*, das sind Wesen, die dem Menschen[Mensch] *sehr treu* sind.

- Entwurf eines typischen Schuhs
  Eine schmale[schmal] Form, der Fuß sieht[sehen] so akkurater aus. Kleiner Absatz, bis 5 cm, ist bequemer[bequem], als ein hoher[hoch] Absatz oder gar keiner. Dunkle Farbe, eher schwarz, denn diese Farbe passt[passen] zu allem, im Gegensatz zu bspw. rot, gelb oder grün. Am Wichtigsten: die Schuhe[Schuh] müssen weich und bequem sein.

  Verkauf Sprosse Alinea
  Hacken Lasche Zunge
  Abschnitt Tritt Niveau Stufe
  Paragraph Leder Markt

- person
  Sehr bequem, nicht für jeden Tag, sondern zum Ausgehen. Sie sind schmal und elegant. Eine Frau, mit der ich abends *etwas unternehmen* Ferse Sohle Sockel Phase angenehme[angenehm] und gut aussehende[aussehend] Person, mit der ich gerne Zeit verbringe[verbringen]. Schuhteil Absatz Passus

Frage: sensorische technik

- sehen
  Viele Schuhe[Schuh] in verschiedenen[verschieden] Farben[Farbe], obwohl ich farbige[farbig] Schuhe[Schuh] nicht mag, trotzdem soll es da Schuhe[Schuh] in allen möglichen[möglich] Farben[Farbe] geben. Nicht nur Schuhe[Schuh] für Frauen[Frau], sondern für alle, denn ich mag mir auch Kinder[Kind]- oder Männerschuhe anschauen.

Fig. 10.2 HTML visualisation of processed text

calculated by comparing the synonym vectors with respect to the number of common synonyms using the Dice coefficient [7]. Alternatively, a cosine co-efficient could have been applied which has been shown to be more sensitive to the relative importance of a word [10].

In addition to the creation of the "normal" graph, it is possible to create a "joined" graph in which nodes with a high[1] similarity are joined. In case there are too many nodes in the graph that have many links to each other, this joining of nodes generally achieves better clustering results.

The tool has functions to display the graph and to export it as a matrix. These functions do not change the derived results; they merely allow for a simple visualisation and make it possible to analyse the data with other tools.

The resulting graph is clustered with the Chinese Whispers Clustering algorithm [3]. Each feature is assigned to a cluster. The resulting data are then exported to a CSV[2] file that can easily be imported into other tools for further processing.

10.4 Exemplary Case Study

In order to illustrate the proposed approach, a study was conducted that analyzed the association of female consumers with the product category "shoes". This product category is assumed to be emotionally laden especially for female consumers, reaching far beyond mere functional aspects. While the comfort of shoes influences the physical well-being of a person, it is more than the mere satisfaction of such physical needs that drives the purchase of shoes.

During the buying process, both physical products as well as brand images provide cues for the activation of associative networks, leading to the purchase or rejection of particular products. Using sentiment analysis to reveal subconscious

---

[1] If the similarity exceeds a given value.
[2] Comma-separated values – http://en.wikipedia.org/wiki/Comma-separated_values

associations helps marketers understand how consumers really feel about the product category and enables them to create effective targeted marketing programs.

The study was conducted in Hamburg, Germany, with 30 women between 23 and 57 years of age elaborating on their associations with shoes in 30-min interviews each. The sample contained a mixture of women with various cultural, educational, and sociodemographic backgrounds. Further, as suggested by Supphellen [11], the sample contained heavy users as well as average and light users. Several different questioning techniques, including the presentation of visual stimuli as well as sensory and projective techniques [12] were applied. This allowed for a comprehensive view on the interviewees' associations regarding shoes and the process of purchasing and wearing shoes.

A total of 1,938 different features could be extracted from the transcribed interviews. Manual coding resulted in 133 and 112 categories for the two raters respectively. Inter-rater-reliability was 65.3% after the exclusion of uncategorized features. Interestingly, inter-rater-reliability was 60.6% for emotional aspects while for rational aspects, it was 66.7%. Thus, the hypothesis of the increased subjectivity in the coding of emotional sentiments (as contrasted to coding of rational aspects) could be confirmed.

The automatic categorization resulted in 185 categories or clusters. 100 of the 148 manually developed categories, i.e. 67.6%, were identical or similar to the automatically developed categories. This figure highlights the high quality and accuracy of the clustering algorithm.

The network analytic examination of the processed data yields an associative network as shown in Fig. 10.3. In order to reduce the complexity and make the network intuitively understandable, only links of strength 7 and above and the respective nodes are shown.

It can be seen that the product category of shoes activates a number of highly emotional associations in the female consumers' minds. The experiential aspects of the purchasing process (dark-colored circles) are highly salient as shown by associations such as "satisfy/please", "wear/try on", "spend time", "discover", "examine", "watch/perceive", "satisfaction/gratification", "enjoy", and "bliss." Simply put: the process of selecting and buying shoes makes female consumers happy and gives them a feeling of deep satisfaction. Service quality and store ambience can therefore be strong differentiating factors for a shoe or shoe store brand.

Additionally, shoes are not only seen as part of a woman's appearance, but they also contribute to her overall well-being and self-confidence. Strong links between associations such as "appearance", "attractive/sexy", "comfort", "grace/beauty", "extravagant/unique", "soul", and "mood" (lined circles) highlight the role of shoes as transformers of a woman's perception of herself. The associative network further reveals the noteworthy effect of high heels on a woman's way of walking: with the felt "extension" of her legs, the interviewed female consumers perceive to walk more gracefully and feel more attractive when wearing high heeled shoes.

A third group of nodes (middle dark circles) comprises aspects of the actual quality of a shoe, including its design, form, material, and colour, which is part of a shoe's signalling function. Finally, associations such as "light/sunny",

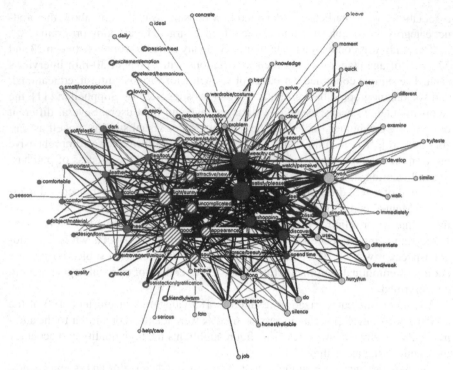

Fig. 10.3 Associative network for shoes

"uncomplicated" and "relaxed/harmonious" help researchers understand how women characterize their relationship with shoes: unlike clothing, shoes fit no matter whether a woman's weight changes, making them very "uncomplicated companions" and the shopping experience very "relaxed."

Translated into marketing activities, shoe brands could gain a significant competitive edge by using a strongly personal and emotional positioning. A communication strategy should be designed that reflects their remarkable transformative effect. Shoes are a highly personal issue, comparable to jewelry, which is why approaches such as mass customization using online design platforms that take a woman's desire for unique shoes into account may have a high potential for future success.

10.5 Conclusion and Outlook

As shown by the case study, automated text analysis offers many interesting opportunities for innovative marketing research applications. The developed tool yields results that are comparable to manual coding of qualitative data while requiring only a fraction of the necessary time and effort. The network representation of the main concepts offers a quick yet comprehensive overview of the complete pool of qualitative data. Further network analytic measures can yield more detailed insights into the roles and relationships of individual associations. Such consecutive analyses, which are beyond the scope of this article, should allow for additional concept disambiguation and a thorough analysis of the interviewees' thoughts and sentiments. This would hardly be possible with manual data processing and purely verbal descriptions of the findings.

Future work can improve the results in various ways. In the field of text analysis, the number of one-element clusters can be reduced. There are clustering algorithms that may perform less powerfully while possibly yielding better results (e. g. agglomerative hierarchical clustering). Also the synonym data is still incomplete. Thus, as of now, some features cannot be clustered because there are no synonyms to be compared. Another task is the extraction of valence words. At present, they are recognized only if they appear in front of features. But human language is more variable. For example consider the following sentence: "They may say it is delicious, but it is not!" To handle wording like this appropriately, more complex patterns will be needed.

On the marketing side, we see a range of opportunities arising from the application of automated text analysis. However, marketing researchers must look beyond the mere extraction and clustering of features. Taking the network structure of human memory into account, possible future research should aim at reconstructing both the order as well as direction of node activation. This would allow for an even deeper understanding of purchasing motivation and decision processes. Additionally, data elicitation and interview transcription techniques should be adjusted to ex-ante accommodate for the specifics of automated text analysis tools in order to yield the most useful and precise data possible.

References

1. Anderson, J.R. 1983. *The architecture of cognition*. Cambridge, MA: Harvard University Press.
2. Biemann, C., S. Bordag, U. Quasthoff, and C. Wolff. 2004. Web services for language resources and language technology applications. In Proceedings Fourth International Conference on Language Resources and Evaluation, LREC 2004.
3. Biemann, C. 2006. Chinese Whispers – an Efficient Graph Clustering Algorithm and its Application to Natural Language Processing Problems. In Proceedings of the HLT-NAACL-06 Workshop on Textgraphs-06, New York.
4. Collins, A.M., and E.F. Loftus. 1975. A spreading-activation theory of semantic processing. *Psychological Review* 82(6):407–428.
5. Henderson, G.R., D. Iacobucci, and B.J. Calder. 2002. Using network analysis to understand brands. *Advances in Consumer Research* 29(1):397–405.
6. Herrmann, M., E. Ruppin, and Usher, M. 1993. A neural model of the dynamic activation of memory. *Biological Cybernetics* 68:455–463.
7. Heyer, G., U. Quasthoff, and T. Wittig. 2006. Text Mining – Wissensrohstoff Text. W3LVerlag, Bochum 2006.
8. Keller, K.L. 1993. Conceptualizing, measuring, and managing customer-based brand equity. *Journal of Marketing* 57(1):1–22.
9. Krishnan, S.H. 1996. Characteristics of memory associations: A consumer-based brand equity perspective. *International Journal of Research in Marketing* 13(4):389–405.

10. Lewis, J., J. Hicks, M. Errami, and H. Garner. 2006. Text similarity: An alternative way to search MEDLINE. *Bioinforatics* 22(18):2298–2304.
11. Supphellen, M. 2000. Understanding core brand equity: Guidelines for in-depth elicitation of brand associations. *International Journal of Marketing Research* 42(3):319–338.
12. Teichert, T., K. Valta, and D. von Weissenfluh. 2004. Entwicklung einer "Marke Bern". In *Markenforschung*, ed. Ch. Zerres, 311–329. Reiner Hampp, München.
13. Zaltman, G., and Coulter, R.H. 1995. Seeing the voice of the customer: Metaphor-based advertising research. *Journal of Advertising Research* 35(4):35–51.

Chapter 11
Automating Opinion Analysis in Film Reviews: The Case of Statistic Versus Linguistic Approach

Damien Poirier, Cécile Bothorel, Émilie Guimier De Neef, and Marc Boullé

11.1 Introduction

With the spread of high speed access to the Internet and new technologies, there has been tremendous growth in online music and video markets. As more players appear on this field, competition increases and content provider can no longer wait for the customer. Instead, they try to trigger purchases by pushing contents: suggesting different choices of movies or songs has become the big thing when it comes to selling content on-line. Actually recommendation is not a new concept, it has already been used on commercial sites (*Amazon, Fnac, Virgin* ...) as well as on musical platforms (*Lastfm, Radioblog, Pandora* ...). But looking at the recommendation techniques used on such web sites suggests that there is still room for innovation.

Candillier et al. [5] presents an overview of recommendation techniques. These techniques are either based on Internet users' notations or content descriptions (*user-* and *item-based* techniques using collaborative filtering), or based on matching Internet user profiles and content descriptions (content filtering), or based on hybrid techniques combining both approaches. Although these techniques are different, they have the same limitation: the hollow nature of a matrix describing users and content profiles. Indeed, the sites proposing recommendations to their customers often have a large catalogue while users only give their opinion on a small number of products. This phenomenon makes the comparisons between profiles risky. In the recommendation field, the difficulty in collecting descriptions about users taste (ratings, interests ...) and content (meta data) is a recurrent problem.

In order to address these problems, a new research area has opened up: mining the resources of the *open* Internet to boost *closed* sites performance. Instead of focusing solely on the data that can be retrieved from a single web site, recommendation techniques may include the vast amount of data that is now available from the Internet. In the era of Web 2.0 and community sites, it is now common for users to share

D. Poirier (✉)
France Telecom RD, TECH/EASY, 22300 Lannion, France; Le Laboratoire d'Informatique Fondamentale d'Orléans, Université d'Orléans, Rue Léonard de Vinci, F-45067 Orleans Cedex 2, France
e-mail: damien.poirier@univ-orleans.fr

K. Ahmad (ed.), *Affective Computing and Sentiment Analysis*, Text, Speech
and Language Technology 45, DOI 10.1007/978-94-007-1757-2_11,
© Springer Science+Business Media B.V. 2011

pictures, tags, news, opinions . . . Such data could be gathered to support automatic information extraction.

Motivated by this potential shift in providing recommendation, we have focused on methods for extracting opinions from movie reviews published on community sites.[1] Our main objective is to establish a user profile based on what he or she declares to like or dislike in movies through his or her published writings (blogs, forums, personal page on the flixster website, etc . . .).

We focus on two different opinion extracting methods. The first machine-learning (ML) method was developed to classify textual reviews into either a positive or negative class. The second natural language processing (NLP) based system is used to build an opinion dictionary and to detect words carrying opinion in the corpus and then to predict an opinion.

We did apply those two approaches to data from the movie-review web-site, www.flixter.com. We discuss the results to compare the two approaches and we provide insights as to which approach should be used for a given corpus of opinions.

11.2 Related Work

Opinion extraction in (trademark) product reviews is important. For instance, [7] present a method for automatically classifying reviews according to the polarity of the expressed opinions, i.e. the tool labels reviews positively or negatively. They index opinion words and establish a scale of rates according to intensity of words. They determine words intensity by using machine learning techniques. Finally, to classify a new review, they build an index reflecting the polarity of each sentence by counting identified words.

[25] explain how they verify reputation of targeted products by analyzing customers' opinions. They start by finding Web pages *talking* about a product, for example a television, then they look for sentences which express opinions in these websites, and finally they determine if the opinions are negative or positive. They determine this by locating in reviews opinion words which were indexed previously in an *opinion dictionary*. Related work of import here includes [31] who have classified reviews in two categories: recommended and not recommended; Wilson et al. [35] which categorizes sentences according to polarity and strength of opinion; and, [26] which seeks opinions on precise subjects in documents.

11.2.1 Machine Learning for Opinion Analysis

Systems using learning machine techniques generally classify textual comments into two classes, positive and negative; extensions of these methods incorporate a

---

[1] This work enters in the frame of European project IST Pharos (PHAROS is an Integrated Project co-financed by the European Union under the Information Society Technologies Programme (6th Framework Programme), Strategic Objective "Search Engines for Audiovisual Content" (2.6.3)).

third class, neutral, and sometimes the classification comprises five classes, very positive, positive, neutral, negative and very negative. These supervised classification methods assume a comment describes only one product and try to predict the rate given by the author.

Many methods use NLP techniques to annotate the corpus. Wilson and Wiebe [34] describe a scheme for annotating expressions of opinions, beliefs, emotions, sentiment and speculation [. . .] in the news and other discourse; Wilson et al. [35] test three different learning methods, frequently used by linguists: *BoosTexter* [29], *Ripper* [6] and *SVM*^{light}, an implementation of Vapnik's [33] Support Vector Machine by [16]. The use of *SVM*^{light} gives the best results on [35] annotated corpus. Pang et al. [27] use a naive Bayes classifier and a classifier maximizing the entropy. Similarly, in order to characterize what is appreciated or not appreciated in a sentence, [37] combine a *parsing* technique with a Bayes classifier to associate polarity with sets of themes.

Furthermore, [27] and [7] show that corpus preparation with a lemmatizer or a negation detection for example, does not lead to better annotation. In order to predict reviewers' opinion, these two papers explore learning methods and show that these methods are more powerful than parsing methods followed by a calculation as shown in the next section. If reviewers' comments are treated as bags of words and a relevant learning technique is used, this leads to 83% correct predictions. We will show that our own experiments confirm these conclusions.

11.2.2 Linguistic Methods of Opinion Analysis

[23] describe their *Opinion Observer* system which compares competitive products by using product reviews left by the Internet users. The system finds features such as pictures, battery, zoom size, etc. in order to explain the sentiment about digital cameras. *Opinion Observer* is a supervised pattern discovery method for automatically identifying product features described in the reviews. The system uses a five step algorithm to analyse reveiwers' comments:

Step 1. PERFORM Part-Of-Speech (POS) tagging and remove digits: For example, the comment "Battery usage; included software could be improved; included 16 MB is stingy" will be transformed as *<N> Battery <N> usage <V> included <N> MB <V>is <Adj> stingy*

Step 2. REPLACE actual feature words in a sentence with *[feature]*: *[feature] <N> usage <V> included <N> [feature] <V> is <Adj> stingy*

Step 3. USE n-gram to produce shorter phrase embedded in a long clause: *<V> included <N> [feature] <V> is <Adj> stingy*, will be parsed into two smaller segments: "*<Adj> included <N> [feature] <V> is* and *<N> [feature] <V> is <Adj> stingy*. (Only 3-grams are used in *Opinion Observer*)

Step 4. DISTINGUISH duplicate tags.

Step 5. PERFORM word stemming:

This five step algorithm generates 3-gram segments which are saved in a transaction file. *Opinion Observer* then uses association rule learning algorithm to extract rules like

(a) <N1>, <N2> –> [feature]
(b) <V>, easy, to –> [feature]
(c) <N1> –> [feature], <N2>
(d) <N1>, [feature] –> <N2>

Not all the generated may help in extracting product features, so *Opinion Observer* selects only the "relevant" rules:

 (i) Rules that have [feature] on the right-hand-side
 (ii) Rules that have the appropriate sequence of items in the conditional part of each rule.
 (iii) Rule Transformation: Rule, as such, still cannot be used to extract features and have to be transformed into patterns to be used to match test reviews.

Moreover, there are other steps that help to group synonyms and to deal with the weighting that has to be attached to opinion-bearing terms. The system then decides the orientation of the extracted feature according to the words extracted near the features. Then the system classifies sentences as negative or positive by determining the dominant orientation of the opinion words of the sentence. The result of the comparison between two products is given in the form of diagram with features on X-coordinate and opinions polarity on Y-coordinate.

Opinion Observer is an example of a system based on the analysis of sentences that facilitates in computing the frequency of sentiment-bearing text excerpts (words, expressions, and patterns). Like many similar systems [25, 26, 31, 35], the *Opinion Observer* needs an *Opinion Dictionary* with as many words or expressions as possible that are used for expressing opinions. To build such a dictionary, different techniques can be used but in almost all the cases there is one proviso: the hand-crafting of a set of words and expressions that are used in expressing an opinion, especially polar or neutral opinion. This set is usually referred to as a *seed* and the aim is to find other words and expressions yielding opinions and classify them according to their semantic orientation (positive, negative, but seldom neutral).

Such an opinion-annotated lexicon can be built by using machine learning techniques. For example, [12] and [32] use an unsupervised learning algorithm to associate new words with the seed words. Pereira et al. [28] and [21] describe methods of discovering synonyms by analyzing words collocation.

Linguistic methods use syntactic and grammatical analyses in order to extend the lexicon. Hatzivassiloglou and McKeown [12] use conjunctions between a word for which semantic orientation is known and an unclassified word. For example, if there is the conjunction *and* between two adjectives, we can consider that the two have the same polarity if any. Contrariwise, if there is the conjunction *but* between two adjectives, then there is a good chance that the two adjectives have a different semantic orientation.

Turney [31] uses more complex patterns and count the frequency of the words, or expressions beside a word, or expression already classified, and define the semantic orientation of those new words or expressions according to the orientation of the neighbours. Each time an adverb or an adjective is encountered, a pair of consecutive words is extracted:

- Adjective with noun
- Adverb with adjective when they are not followed by a noun
- Adjective with adjective when they are not followed by a noun
- Noun with adjective when they are not followed by a noun
- Adverb with verb

The second extracted word allows the system to confirm polarity of the adjective or adverb by giving an outline of the sentence's context.

The above mentioned method, of counting co-occurrences with words semantically oriented and manually selected, is also used in [36] to determine words are semantically oriented, in terms of the direction and the strength of the orientation. To measure more precisely the strength of opinion expressed in a sentence, adverbs which are associated to adjectives are extracted. Indeed, [1] propose a classification of adverbs into five categories: adverbs of affirmation, adverbs of doubt, adverbs of weak intensity, adverbs of strong intensity and adverbs which have a role of minimizer. The strengths of adverb-adjective combinations are computed according to the weights assigned to these five different adverb categories.

Google's work [10] attempts to find semantic orientation of new words from *WordNet* databases [24]. In similar vein, [14] use sets of synonyms and antonyms present in WordNet to predict semantic orientation of adjectives. In WordNet, words are organised in tree (see Fig. 11.1). To determine polarity of a word, the system traverses the trees of synonyms and antonyms of this word and if it finds a seed

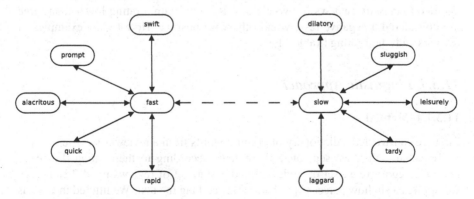

Fig. 11.1 Tree of synonyms and antonyms in WordNet (*full arrow* = synonyms, *dotted arrow* = antonyms)

word in the synonyms, it allocates the same class, but if it finds a seed word in the antonyms, it allocates the opposite class. If it does not find a seed word, it rebuilds the analysis with synonyms and antonyms, and loops until a seed is found.

In our opinion, method outlined in [10] is not well-grounded as words can have different meaning according to the context and thus can have synonyms not signaling the same thing. For example, the word *like* has for synonym *love*, but in the sentence *It is like that*, one can use the synonym *love* instead of *like* for instance.

To associate a polarity, negative or positive, to a sentence, we can count the number of terms with positive semantic orientation and the number of terms with negative orientation. If there are more positive terms, the sentence is declared positive, if there are more negative terms, the sentence is declared negative, and if there are as many positive as negative terms, either sentence is declared neutral [36]; with another strategy, the last term carrying opinion determines the sentence polarity [14]. Otherwise, we can extract opinion one by one associated with the feature it refers to [14, 35].

11.3 Linguistic and Machine Learning Methods: A Comparative Study

In this section we compare two opinion analysis methods using reviews rated by authors and their results. These reviews are available as a corpus of texts. The initial corpus comprises 60,000 films divided equally in positive and negative reviews. We have use 50,000 reviews for training and 10,000 for testing.

The main difficulty with this corpus is the small size of reviews (12 words on average). This makes opinion extraction difficult, even for humans. Moreover, the corpus comprises textual messages very similar to forum messages and include punctuation marks "!!!", emoticons ":-)", expressions from SMS texts *"ur"*, *"gr8"* and word equivalent for emphasis (*veryyyyy cooooool* instead of *very, very cool*).

Each review in our corpus has a rating given by the author, on a scale of 0 (zero) to 5, and our aim is to predict this rating from our automatic analysis. We have decided to classify reviews in two classes. Reviews with a rating lower than three are considered a negative review and otherwise positive. Here follow examples of reviews with their rating (Table 11.1).

11.3.1 Linguistic Approach

11.3.1.1 Method

First we constructed a dictionary of opinion words from a reviewer's corpus.

To achieve this, we separated all reviews according to their rating. For each review category (e.g., set of reviews rated 1 star, set of reviews rated 2 stars . . .), we applied a shallow parser [8] to lemmatize and tag the text. We filtered the words according to their Part of Speech tag and frequency. Verbs and adjectives have then been manually classified according to the opinion they convey.

Table 11.1 Polarity (NEG/POS) and exemplar reviews

Rate	Review
POS	Great movie!
NEG	This wasn't really scary at all i liked it but just wasn't scary...
POS	I loved it it was awesome!
NEG	I didn't like how they cursed in it......and this is suppose to be for little kids....
NEG	Sad ending really gay
POS	sooo awesome!! (he's soo hot)
POS	This is my future husband lol (orlando bloom)
NEG	Will Smith punches an alien in the face, wtf!!??
NEG	I think this is one of those movies you either love or hate, i hated it! :o)

Table 11.2 Part of hand-crafted lexicon

Positive words	Good, great, funny, awesome, cool, brilliant, hilarious, favourite, well, hot, excellent, beautiful, fantastic, cute, sweet...
Negative words	Bad, stupid, fake, wrong, poor, ugly, silly, suck, atrocious, abominable, awful, lamentable, crappy, incompetent...

This list has been expanded using a synonym dictionary (www.wordreference. com). Only verbs and adjectives that are not ambiguous have been classified. For example, the word *terrible* is not classified because it can expresses both opinion polarities.

A total of 183 opinion words have been classified in two classes, positive words (115) and negative words (68), in this manner. An excerpt from the dictionary is show in Table 11.2. This dictionary was not made using the corpus used to evaluate this method.

The last step of the analysis consists of counting opinion words in each review to determine the polarity. For that we first lemmatized all reviews (the same pretreatment as the lexicon) and only adjectives and verbs were kept. Then, a polarity was assigned to reviews according to the majority number of positive words or negative words.

No sophisticated NLP techniques were performed, such as a grammatical structural analysis. But keeping only verbs and adjectives avoids misinterpretations of words such as *"like"* which can have different roles in a sentence.

11.3.1.2 Results

This method enabled us to rate 74% of film reviews on the 20,000 present in the test corpus. All the following results are calculated according to the rated reviews. To compare results with other techniques, we calculate three values: precision, recall and F_{score}.

Table 11.3 Confusion matrix obtained with the hand-crafted lexicon		Pos. rev.	Neg. rev.
	Predicted pos. reviews	8,089	3,682
	Predicted neg. reviews	218	2,823

- $precision = \dfrac{number\ of\ positive\ examples\ cover}{number\ of\ examples\ cover}$

- $recall = \dfrac{number\ of\ positive\ examples\ cover}{number\ of\ positive\ examples}$

- $F_{score} = \dfrac{2 * precision * recall}{precision + recall}$

The confusion matrix of results is presented in Table 11.3.

With our technique we obtained 0.81 for precision, 0.70 for recall and 0.75 for F_{score}.

Our principal problems was in determining the polarity of negative reviews. Indeed, the recall of negative reviews is 0.43, whereas it is 0.97 for positive reviews. Contrarily, precision of positive reviews (0.69) is worse than precision of negative reviews (0.93).

This problem can be related to our dictionary: the positive category contains almost twice as many words than negative category. However, the problem is not the detection of negative reviews but their interpretation. These results lead us to think that sometimes people use negation of a positive expression to express their negative feelings without using an adjective or verb carrying negative opinion. This intuition will be confirmed by using a statistical method.

In order to evaluate the quality of our dictionary, an experiment was performed using a set of English words already classified by [30] and [17]. The new lexicon contains 4,210 opinions words (2,293 negative words and 1,914 positive words). With this new opinion dictionary, the technique classifies more reviews (a gain of 4% essentially on negative ones) but prediction results are worse than previous: 0.67 for precision, 0.65 for recall and 0.66 for F_{score}. See Table 11.4.

The explanation for these results is certainly a lexicon less adapted to this corpus. It is a more general lexicon whereas our lexicon was built with words appearing regularly in a similar corpus.

These new results show the same problem with negative reviews, though this second lexicon contains more negative words. This confirms our first idea that negation is an important part of better interpretation of negative reviews.

Table 11.4 Confusion matrix obtained with *General Inquirer* lexicon		Pos. rev.	Neg. rev.
	Predicted pos. reviews	7,027	3,743
	Predicted neg. reviews	1,165	3,716

11.3.1.3 Analyzing Errors

There are errors that we wish to discuss in particular - one associated with review that our method could not rate and the other error was associated with poorly rated reviews.

Unrated reviews

There are several explanations why reviews are not rated:

- Gaps in the hand crafted lexicon. Examples: "wooohooo film", "watched it all the time when i was younger", "no please no", "I can not remember story".
- Presence of adjectives expressing sentiments or beliefs that can be associated with different opinion. Examples: "so romantic", "weird movie", "I was afraid", "it is very sad".
- Presence of as many positive words as negative words. In this case, the classifier considers the review neutral. Examples: "<u>bad</u> dish <u>good</u> opinion", "not <u>bad</u> - not <u>great</u> either", "really <u>bad</u> film, I thought it would be a <u>lot better</u>".
- Some of the reviews are emptied by the NLP pretreatment. They don't contain any verbs or adjectives.

Poorly rated reviews

The majority of errors are due to negation words which are not considered in this approach. The solution could be to change opinion polarity when a negation is present in the review. Indeed, reviews are very short, statistically, that is, these reviews are composed of only one sentence, thus the negation modifies the polarity of all the verbs or adjectives present. Perhaps what is needed is dependency parsing in order to find which word the negation is related to, and thus reversing the polarity only on the involved words.

We can find numbers of ironic or sarcastic sentences as "fun 4 little boys like action heroes and stuff u can get into it :p" which was rated negatively by the author whereas we rate it positively.

11.3.2 Machine Learning Approach

Let us first present the method we used and then comment on the results. We will analyze the prediction quality of our classifier, and we will show how a deeper exploration gives information on the Internet users' writing style.

11.3.2.1 Compression-Based Averaging of Selective Naive Bayes Classifiers

In this section, we summarize the principles of the method used in the experiments. This method, introduced in [3], extends the naive Bayes classifier owing to optimal preprocessing of the input data, to an efficient selection of the variables and to averaging the models.

Optimal discretization

The naive Bayes classifier has proved to be very effective on many real data applications [11, 19]. It is based on the assumption that the variables are independent in each output label, and relies on an estimation of univariate conditional probabilities.

The evaluation of the probabilities for numeric variables has already been discussed in the literature [9, 22]. Experiments demonstrate that even a simple equal width discretization brings superior performance compared to using a Gaussian distribution.

In the MODL approach [2], the discretization is turned into a model selection problem. First, a space of discretization models is defined. The parameters of a specific discretization are the number of intervals, the bounds of the intervals and the output frequencies in each interval. Then, a prior distribution is proposed on this model space. This exploits the hierarchy of the parameters: the number of intervals is first chosen, then the bounds of the intervals and finally the output frequencies. The choice is uniform at each stage of the hierarchy.

Finally, the multinomial distributions of the output values in each interval are assumed to be independent. A Bayesian analysis is applied to select the best discretization model, which is found by maximizing the probability $p(Model|Data)$ of the model given the data.

Owing to the definition of the model space and its prior distribution, the Bayes formula is applicable to derive an exact analytical criterion to evaluate the posterior probability of a discretization model.

Efficient search heuristics allow us to build the most probable discretization given the data sample. Extensive comparative experiments report high performance.

Bayesian Approach for Variable Selection

The naive independence assumption can lead to misleading inference when the constraints are not respected. In order to better deal with highly correlated variables, the selective naive Bayes approach [20] uses a wrapper approach [18] to select the subset of variables which optimizes the classification accuracy.

Although the selective naive Bayes approach performs quite well on datasets with a reasonable number of variables, it does not scale on very large datasets with hundreds of thousands of instances and thousands of variables, such as in marketing applications or, in our case, text mining. The problem comes from the search algorithm, whose complexity is quadratic in the number of the variables, and from the selection process which is prone to over fitting.

In [3], the overfitting problem is solved by relying on a Bayesian approach, where the best model is found by maximizing the probability of the model given the data.

The parameters of a variable selection model are the number of selected variables and the subset of variables. A hierarchic prior is considered, by first choosing the number of selected variables and then choosing the subset of selected variables. The conditional likelihood of the models exploits the naive Bayes assumption, which directly provides the conditional probability of each label. This allows an exact calculation of the posterior probability of the models.

Efficient search heuristics with super-linear computation time are proposed, on the basis of greedy forward addition and backward elimination of variables.

Compression-Based Model averaging

Model averaging has been successfully used in Bagging [4] with multiple classifiers trained from re-sampled datasets. In this approach, the averaged classifier uses a voting rule to classify new instances. Unlike this approach, where each classifier has the same weight, the Bayesian Model Averaging (BMA) approach [13] weights the classifiers according to their posterior probability.

In the case of the selective naive Bayes classifier, an inspection of the optimized models reveals that their posterior distribution is so sharply peaked that averaging them according to the BMA approach almost reduces to the MAP model. In this situation, averaging is useless.

In order to find a trade-off between equal weights as in bagging and extremely unbalanced weights as in the BMA approach, a logarithmic smoothing of the posterior distribution called compression-based model averaging (CMA) is introduced in [3].

Extensive experiments have demonstrated that the resulting compression-based model averaging scheme clearly outperforms the Bayesian model averaging scheme.

11.3.2.2 Results

Recall that in a machine learning approach there is no a priori data: We used the original reviews by the authors; all tokens were rendered into lower case letters and all punctuation marks were removed. The reviews were processed as a bag of words. We trained our system on a corpus containing 20,000 positive reviews and 20,000 negative. We then used a test corpus to evaluate the training regimen.

The training corpus comprised 24,825 tokens: Our system found only 305 tokens to be informative. Very few of the tokens are very informative (Fig. 11.2) and are

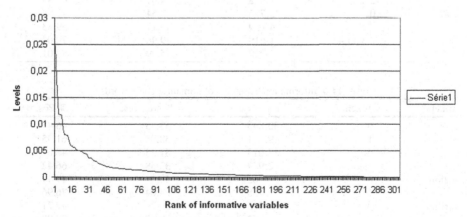

Fig. 11.2 Evolution of levels of informative variables

classified according an associated *level* value. The *level* value is directly related to the posterior probability of a discretization model, with a 0–1 normalization. The zero level indicates that token has no information, and the unity value (the value only approaches unity asymptotically) suggests that the token has maximum information.

The majority of words having a positive level express opinion, but other words also appear in this list.

We found that some of the grammatical words that have a very high frequency in general language texts are comparatively rare in our movie review corpus: over 70% of the reviews do not have determiners *a, the* together with the conjunction *and*. It turns out that the presence of relatively "rare" words in the reviews that do contain these grammatical words have a positive polarity. (See Table 11.5).

The occurrence of the domain-specific words, for example *film, movie* and qualifiers *action, thriller*, are comparatively rare in our corpus: The more formal *film* occurs only in 8% of the reviews and the informal *movie* in around 23% of the reviews. However, it appears that the reviewers who write positive reviews use these terms in around 2 in 3 of all those reviews comprising these domain words. The even rare qualifiers, *action* and *thriller* appearing in no more than 1 and 2% respectively of all the 40,000 reviews, are used in 75% (and 90% respectively) of the positive reviews that have the qualifiers. (See Table 11.6).

Intuitively, negation words appear mainly in negative polarity reviews despite the fact that two of these types of words *didn't* and *not* appear only in 3 and 10% of all our reviews (Table 11.7). This explains the weak score of precision for positive

Table 11.5 Information "content" of grammatical tokens *a, and* and *the*

Token	Token f	Tot. reviews	% Neg. rev.	% Pos. rev.
a	0	31,177	52.31	47.69
	1	7,960	43.05	56.95
	2	863	30.48	69.52
the	0	29,179	52.29	47.71
	1	8,923	45.77	54.23
	2	1,898	34.61	65.39
and	0	33,725	52.54	47.46
	1	4,439	40.17	59.83
	2,3, or 4	1,619	29.96	70.04
	5	217	5.99	94.01

Table 11.6 Information "content" of domain specific words

Token	Token f	Tot. reviews	% Neg. rev.	% Pos. rev.
movie	0	30,849	53.55	46.45
	1	9,151	38.05	61.95
film	0	37,013	51.33	48.67
	1	2,987	33.58	66.42
action	0	39,262	50.51	49.49
	1	738	23.04	76.96
thriller	0	39,725	50.29	49.71
	1	275	8.00	92.00

Table 11.7 Information
content of *not* and *didn't*

Token	Token f	Tot. reviews	% Neg. rev.	% Pos. rev.
not	0	36,189	48.51	51.49
	1	3,811	64.13	35.87
did'nt	0	38,896	49.42	50.58
	1	1,104	70.47	29.53

Table 11.8 Confusion matrix
obtained with machine
learning

	Pos. rev.	Neg. rev.
Pos. reviews predict	7,060	1,793
Neg. reviews predict	2,940	8,207

reviews and recall for negative reviews in the previously approach. Indeed, we can note that negation terms appear much more in negative reviews than in positive reviews.

Concerning the opinion prediction, the confusion matrix of results in Table 11.8 shows that this time, all the reviews are classified. Scores obtained are 0.77 for precision, 0.76 for recall and F_{score}. They are better than those obtained with the classic naive Bayes classifier (approximately 0.70 for the three indicators). Results are equivalent to our linguistic results regarding to the F_{score}, but, recall is significantly better for negative reviews (0.82 instead of 0.43), as is the precision on positive reviews (0.80 instead of 0.69). On the contrary, recall is worse for positive reviews (0.70 instead of 0.97) and so is the precision on negative reviews (0.74 instead of 0.93). The ML technique provides balanced results for each class, but overall it does not outperform the NLP approach.

11.4 Conclusion and Prospects

We have tested and evaluated two approaches for opinion extraction. The first one consists of building a lexicon containing opinion words using *low-level* NLP techniques. This lexicon facilitates the classification of reviews as either positive or negative. The second method consists of using a machine learning technique to predict the polarity of each review.

We used data from the flixster website as a benchmark to evaluate those two recommendation methods, using part of the opinion corpus as a learning testbed and the rest of it to evaluate classification performance: we were are able to discriminate the qualities of the two techniques according to various criteria. In the rest of this conclusion, we synthesize our results, trying to provide the reader with an understanding of each technique's specificity and limitation.

The results obtained with the machine learning (ML) technique appear to provide an inherently deeper understanding of how the authors express themselves according to what they thought about a movie. Indeed, they show that people generally write more when they appreciated the movie for example, giving more detailed reviews of movies features. It turns out that opinion words are not the only opinion indicator, at least for this kind of corpus.

Independently of the analysis technique, an important issue with automating opinion extraction is that we cannot expect a machine to predict good polarity for each review. Consider for instance the sentence "*Di Caprio is my future husband*": it does not indicate whether the author appreciated the film or not. Thus our aim is not to know the polarity of each review but to have the best possible classification. Improvement of prediction results with ML can be obtained by using an indecision threshold. i.e. when the probability for a good prediction is too weak, we can decide not to classify the review.

With the NLP technique, this problem does not exist because reviews which do not contain opinion words are not classified. However, results from this technique can be improved. For instance, detecting negations would be an important step forward. Indeed, ML results show that negative opinions are often expressed by using words carrying positive opinion associated with a negation. Since our linguistic approach ignores every negation, most of the negative reviews are labeled as positive ones. The best solution is probably to proceed to a dependency parsing. But the kind of prose we are faced with (SMS writing, spelling errors, strange sentence construction . . .) will complicate this step.

The main advantage of the ML technique is that new datasets can be analysed without a priori knowledge (i.e. lexicon) and then be deployed with a confidence for both positive and negative reviews. However, the corpus has to be large enough to offer a consistent training dataset and must contain ratings to supervise the training.

This approach may also be used to detect pertinent words and help build dictionary, particularly in the context of Web Opinion Mining, where it is necessary to adapt the lexicon to the *inventive* vocabulary of Internet users' writings.

Contrarily, NLP techniques do not require a learning step, except regular updates to the lexicon. So it can be deployed immediately on a small corpus without rated examples. With a dependency parsing step in order to detect negations, the results could be competitive with ML techniques.

By way of a conclusion, we propose to using a *low-level* NLP approach when the corpus is too small to facilitate good training: the cost of building a lexicon (small ones bring satisfying quality) and designing a negation detection remains reasonable. If the corpus is large enough, ML approaches will be easier to deploy.

To go further, we may explore whether linguistic pretreatments on the corpus for ML techniques can reduce the number of variables (by reducing the vocabulary describing the reviews) without losing information and damaging the quality. We may also focus on a higher level NLP approach and try to explain why people (dis)like movies.

References

1. Benamara, Farah, Carmine Cesarano, Antonio Picariello, Diego Reforgiato, and VS Subrahmanian. 2007. Sentiment analysis: Adjectives and adverbs are better then adjectivesalone. Proc. Int. Conf. on Weblogs and Social Media (ICWSM), (available at http://www.icwsm. org/papers/paper31.html, seen 24/02/2011).

2. Boullé, M. 2006. MODL: a Bayes optimal discretization method for continuous attributes. Machine Learning, 65(1):131–165.
3. Boullé, M. Compression-based averaging of selective naive Bayes classifiers. Journal of Machine Learning Research, 8:1659–1685, 2007.
4. Breiman, L. 1996. Bagging predictors. Machine Learning, 24(2):123–140.
5. Candillier, Laurent, Frank Meyer, and M. Boullé. 2007. Comparing state-of-the-art collaborative filtering systems. International Conference on Machine Learning and Data Mining MLDM 2007, Leipzig/Germany.
6. Cohen, William W. 1996. Learning trees and rules with set-valued features. Proc. AAAI, 709–716.
7. Dave, Kushal, Steve Lawrence, and David M. Pennock. 2003. Mining the peanut gallery: Opinion extraction and semantic classification of product reviews. WWW '03 Proceedings 12th Int. Conference on World Wide Web. (available at http://citeseerx.ist.psu.edu/viewdoc/download?doi=10.1.1.13.2424&rep=rep1&type=pdf) 2003.
8. Guimier de Neef, E., M. Boualem, C. Chardenon, P. Filoche, and J. Vinesse. 2002. Natural language processing software tools and linguistic data developed by France Telecom R&D. Indo European Conference on Multilingual Technologies, Pune, India.
9. Dougherty, J., R. Kohavi, and M. Sahami. 1995. Supervised and unsupervised discretization of continuous features. In Proceedings of the 12th Int. Conf. on Machine Learning, 194–202. Morgan Kaufmann, San Francisco, CA, 1995.
10. Godbole, Namrata, Manjunath Srinivasaiah, and Steven Skiena. 2007. Large-scale sentiment analysis for news and blogs. Proc. Int. Conf. on Weblogs and Social Media (ICWSM), 2007. Available at http://www.icwsm.org/papers/paper26.html, seen 24/02/2011.
11. Hand, D.J., and K. Yu. 2001. Idiot Bayes? Not so stupid after all? *International Statistical Review* 69(3):385–399.
12. Hatzivassiloglou, Vasileios, and Kathleen R. McKeown. Predicting the semantic orientation of adjectives, 1997. In ACL '98 Proceedings of the 35th Annual Meeting of the Association for Computational Linguistics (ACL) and 8th Conference of the European Chapter of the ACL, 174–181. doi:10.3115/979617.979640.
13. J.A. Hoeting, D. Madigan, A.E. Raftery, and C.T. Volinsky. 1999. Bayesian model averaging: A tutorial. Statistical Science, 14(4):382–417.
14. Hu, Minqing, and Bing Liu. Mining and summarizing customer reviews, 2004. Proc. of the 10th ACM SIGKDD Int. Conf. on Knowledge Discovery and Data Mining (KDD '04), 168–177. Seattle, WA, USA – 22–25 August 2004. doi:10.1145/1014052.1014073.
15. Hu, Minqing, and Bing Liu. 2004. (Mining opinion features in customer reviews, In *AAAI'04 Proceedings of the 19th Nat. Conf. on Artificial Intelligence*, ed. A. G. Cohn, 755–760. Available at https://www.aaai.org/Papers/AAAI/2004/AAAI04-119.pdf.
16. Joachims, Thorsten. Making large-scale support vector machine learning practical Thorsten Joachims, 1999. In *Advances in Kernel methods: Support vector learning*, eds. Bernhard Schölkopf, Christopher J.C. Burges, and Alexander J. Smola, 169–184. Cambridge, MA: MIT Press.
17. Edward, Kelly, and Philip Stone. 1975. Computer recognition of English word senses, North Holland Publishers.
18. Kohavi, R., and G. John. 1997. Wrappers for feature selection. *Artificial Intelligence* 97(1–2):273–324.
19. Langley, P., W. Iba, and K. Thompson. 1992. An analysis of Bayesian classifiers. In *AAAI–10, 10th National Conf. on Artificial Intelligence*, 223–228. San Jose: AAAI Press.
20. Langley, P., and S. Sage. 1994. Induction of selective Bayesian classifiers. In *Proceedings of the 10th Conference on Uncertainty in Artificial Intelligence*, 399–406. Morgan Kaufmann.
21. Lin, Dekang. Automatic retrieval and clustering of similar words, 1998. In COLING '98: Proceedings of the 17th Int. Conf. on Computational linguistics - Volume 2 Association for Computational Linguistics, 768–774, Stroudsburg, PA. doi:10.3115/980691.980696.
22. Liu, H., F. Hussain, C.L. Tan, and M. Dash. 2002. Discretization: An enabling technique. Data Mining and Knowledge Discovery, 4(6):393–423.

23. Liu, Bing, Minqing Hu, and Junsheng Cheng. Opinion Observer: Analyzing and Comparing Opinions on the Web. 2005 In WWW 2005, May 10-14, 2005, Chiba, Japan. Available at http://citeseerx.ist.psu.edu/viewdoc/download?doi=10.1.1.81.7520&rep=rep1&type=pdf.

24. Miller, George A., Richard Beckwith, Christiane Fellbaum, Derek Gross, and Katherine Miller. 1990. Introduction toWordnet: An on-line lexical database. *International Journal of Lexicography* 3(4):235–244.

25. Morinaga, Satoshi, Kenji Yamanishi, Kenji Tateishi, and Toshikazu Fukushima. Mining product reputations on the web, 2002. In Proc. of the 8th ACM SIGKDD Int. Conf. on Knowledge Discovery and Data Mining, 341–349. New York: ACM.

26. Nasukawa, Tetsuya, and Jeonghee Yi. 2003. Sentiment analysis: Capturing favorability using natural language processing. In K-CAP '03 Proc. of the 2nd Int. Conf. on Knowledge Capture, 70–77.

27. Pang, Bo, Lillian Lee, and Shivakumar Vaithyanathan. Techniques Thumbs up? Sentiment Classification using Machine Learning Techniques, 2002. In Proceedings of the Conference on Empirical Methods in Natural Language Processing (EMNLP), 79–86, Philadelphia, July 2002.

28. Fernando Pereira, Naftali Tishby, and Lillian Lee. Distributional clustering of English words, 1994. In ACL '93 Proceedings of the 31st Annual Meeting of Association for Computational Linguistics, 183–190. doi:10.3115/981574.981598.

29. Shapire, Robert E., and Yoram Singer. 2000. Boostexter: A boosting-based system for text categorization.

30. Stone, Philip J., Dexter C. Dunphy, Marshall S. Smith, Daniel M. Ogilvie, and associates. 1996. The General Inquirer: A computer approach to content analysis. Cambridge, MA: MIT Press.

31. Turney, Peter D. 2002. Thumbs Up or Thumbs Down? Semantic Orientation Applied to Unsupervised Classification of Reviews. In Proceedings of the 40th Annual Meeting of the Association for Computational Linguistics (ACL'02), 417–424, Philadelphia, Pennsylvania, 8–10 July 2002. Available as National Research Council of Canada Report No. 44946 at http://nparc.cisti-icist.nrc-cnrc.gc.ca/npsi/ctrl?action=rtdoc&an=8914166.

32. Turney, Peter D., and Michael L. Littman. 1995. Unsupervised learning of semantic orientation from a hundred billion-word corpus, 2004. Available at http://arxiv.org/ftp/cs/papers/0212/0212012.pdf.

33. Vapnik, Vladimir. 2003. The Nature of Statistical Learning Theory. Springer.

34. Wilson, Theresa, and Janyce Wiebe. Annotating Opinions in the World Press. In 4th SIGdial Workshop on Discourse and Dialogue (SIGdial-03). ACL SIGdial. Available at http://www.cs.pitt.edu/wiebe/pubs/papers/sigdial03FixedLater.pdf, site visited 7 February 2011.

35. Wilson, Theresa, Janyce Wiebe, and Rebecca Hwa, 2004. Just how mad are you? finding strong and weak opinion clauses. In *AAAI'04 Proceedings of the 19th Nat. Conf. on Artificial Intelligence*, ed. A. G. Cohn, 761–767. Available – http://www.aaai.org/Papers/AAAI/2004/AAAI04-120.pdf, site visited 7 February 2011.

36. Yu, Hong, and Vasileios Hatzivassiloglou. 2003. Towards answering opinion questions: Separating facts from opinions and identifying the polarity of opinion sentences, 2003. In Proceeding EMNLP '03 Proceedings of the 2003 conference on Empirical methods in natural language processing Association for Computational Linguistics, 129–136.

37. Nigam, Kamal, and Matthew Hurst. 2004. Towards a Robust Metric of Opinion. In *AAAI spring symposium on exploring attitude and affect in text*.

Afterword: 'The Fire Sermon'

Yorick Wilks

This book attempts, boldly in my view, to link two complex and difficult areas currently addressed by artificial intelligence (AI) and natural language processing (NLP) research: metaphor and emotion. Eliot wrote, famously [6], of "mixing memory and desire" and we are after something similar, but in a practical and less poetic way. It may be worth asking what our two target concepts have to do with each other, and what bridge concepts there may be between them, such as computational explications of beliefs and goals. Taking things head on, as it were, it is obvious that many metaphors evoke emotion, as in poetry where that is precisely their intention, that many emotions are normally expressed metaphorically, and many such metaphors are "frozen" in our culture, seeming barely metaphorical at all, such as the "pain" of love and its frustrations.

But the "head on" approach may not be the best one: on the one hand, the fact that many metaphors concern emotion may not tell us much about emotion itself; it is a frequent discussion point about the metaphor work of Barnden [4] that he opts to discuss and model metaphors almost entirely of the mind. Is that or is it not work on metaphor? One might say it is only so if exploring those metaphors tells us something about metaphor in general; otherwise, it is work about mind. On the other hand, emotion cannot require metaphor for its expression: it is just one of the vehicles used to express it. Since Darwin [5], there has been a realization that much or most emotion is expressed by the face and body language, and is shared with (non-verbal) animals. For Darwin emotion is not fundamentally a verbal notion at all, even though, as NLP researchers, that may be our focus in this book.

There are two main traditions of emotion research that have attracted the attention of computational modelers: the first descends from Ortony [12] down to contemporaries like Gratch and Marsella [7] and concerns the relationship of emotions to goals. Ortony actually studied both these key notions in an original way, metaphor and emotion, but, so far as I can see [13], without linking them fundamentally: he was concerned with the way metaphors can convey emotion, rather like Barnden and the mind. In this tradition, goals are modeled and emotion can arise from the

Y. Wilks (✉)
University of Sheffield and Oxford Internet Institute, Sheffield, UK
e-mail: yorick@dcs.shef.ac.uk

K. Ahmad (ed.), *Affective Computing and Sentiment Analysis*, Text, Speech
and Language Technology 45, DOI 10.1007/978-94-007-1757-2,
© Springer Science+Business Media B.V. 2011

success and failure of goal-driven behaviours: e.g. negative emotion from the failure or frustration of goal-seeking.

In the more recent work of Gratch and Marsalla (ibid.) they extend the notion of goal-related emotion towards belief, describing scenarios where a model might change its beliefs about the importance of a goal, thus reducing its distress. This is an interesting linking of emotion to goals via the concept of belief, a notion we shall return to below. This tradition of work on emotion, the goal-related one, is not necessarily confined to models of humans, but covers any goal-seeking entity, whereas the other main NLP tradition (see, for example, Wiebe [16]) is essentially linked to the presence of certain lexical items in speech or writing and thus closer to NLP and the modeling of humans.

This lexical tradition, of determining the emotion or, more broadly, sentiment, content of texts from their lexical content alone, has many strands, some of them much older than AI/NLP itself. It may be worth mentioning here the curious relationship of NLP to Content Analysis (CA, [8]) and its associated *General Inquirer* [15]. This last came from a psychological tradition—Stone was at the Harvard Psychology Department – and developed wholly out of contact with the central tradition of NLP. It has now become widely used in finance and other areas associated to publishing and psychometrics.

When this work started, the prevailing notion of content in AI and NLP was in terms of logical or linguistic structure and so there could be little or no communication with the line of work Stone started. I met him in the 1970s but I believe there was virtually no contact at all between the CA and NLP lines of work. This ignorance of what was going on in psychology went further: it included the successful automatic text grading work of Landauer (e.g. [9]). This undoubtedly came, in part, from a deep belief in AI that AI gave models to (cognitive) psychology and not vice versa! Curiously, that belief did not exclude Miller's [11] work on WordNet, but Miller had always kept very close relations with AI and linguistics and the others had not. After the "empirical turn" in CL/NLP about 1990 and the rise of data-driven and statistical methods, there could be no obvious barrier between CA, and other such methodologies, and CL/NLP. But Ken Litkowski is one of the few people in NLP to take note of CA, though Krippendorf's "alpha" measure has now been adopted and acknowledged (see e.g. [14]).

Ahmad's work on sentiment analysis (e.g. [2] and this book) has certainly straddled the NLP/CA line. This historical breach is now healed: CA has primacy, but NLP has contributed machine learning methodology. One might add that sentiment/emotion tagging of texts is now just one more annotation scheme among many others, and many applications of it can be found on the extensive emotion website[1] of the HUMAINE project.

Another quite different strand of work was an attempt to see a range of phenomena as point-of-view or perspective phenomena: this work began with the modeling of belief [3] in a system called VIEWGEN, which had a number of implementations.

---

[1] www.emotion-research.net

Its underlying principle was that belief states of individuals could be modeled (by other individuals) in terms of default transfer of beliefs between belief spaces corresponding to people (and to things or states as objects of belief); it was a method for calculating such states on the fly, rather than assuming them known, as in the classic belief work in AI of Allen and others [1].

The extensions of this work relevant to the present discussion were, first, the extension of the *Viewgen* spaces by Lee [10] to include goals and desires – thus providing a recursive model of goals and beliefs of the sort that Gratch and Marsella required to support their more recent model of emotions as a side-effect. Secondly, the system had been extended [17] to consider object-to-object space mappings (analogous to those of person-to-person space mappings as belief ascriptions of one person to another) as capturing some aspects of metaphor, seen as property ascriptions from one object to another, as in classic cases like "Smith is vermin" where rat-like features are attributed to Smith.

The underlying claim in this work was that metaphor, belief (and the intensional identification of objects, such as *Mary-seen-as-Fred's-Aunt*) could all be modeled by an underlying mechanism of recursive ascription of properties or propositions across notional space boundaries. The paradigm was never fully evaluated, whatever that might be like in this field, and I mention it here only in the context of efforts to create a unified structural theory to link concepts as diverse as metaphor and emotion. In both the cases just mentioned, and the recent explorations of Gratch and Marsalla, mentioned earlier, it seems that belief may provide the linking concept between them. Perhaps this area of NLP/AI, like Physics, still awaits a real Unified Field Theory[2] to link apparently independent phenomena.

These are no rarified, distant, matters in AI: it is easy to see the pressing need for such unification for agent modeling. Suppose an agent says (metaphorically) "Smith is vermin" and the receiving agent has a goal ELIMINATE-VERMIN as well as DO-NO-HARM-TO-PEOPLE, and that it has some emotion tagging engine that says that "vermin" is a highly negative concept as well as the conflict of not being able to fulfill both these goals (eliminating and not eliminating *Smith*), which will be emotionally frustrating for the agent concerned. Then one can see the need for some degree of unified theory is essential to determine the agent's next move. I have directed the COMPANIONS project, a large EU project on long-term Companion agents that try to wrestle with these issues, and its website[3] includes a simple demo of a Health and Fitness Companion that contains some simple version of both approaches to emotion (tagging of input and goal-driven emotion derivation). This book is a step towards a serious investigation of the unification of cognitive and linguistic function that practical performing systems need very badly.

---

[2] en.wikipedia.org/wiki/Grand_unified_field_theory
[3] www.companions-project.org

References

1. Allen, J. F., and C. R. Perrault. 1980. Analyzing intention in utterances. *Artificial Intelligence* 15(3):143–178.
2. Almas, Y., and K. Ahmad. 2007. A note on extracting 'sentiments' in financial news in English, Arabic & Urdu. In Proceedings of the Second Workshop on Computational Approaches to Arabic Script-based Languages, Linguistic Society of America 2007, Linguistic Institute, Stanford University, Stanford, California. Linguistic Society of America, 1–12.
3. Ballim, A., and Y. Wilks. 1991. Artificial believers: *The ascription of belief*. Erlbaum: Hillsdale, NJ.
4. Barnden, J. A. 2007. Metaphor, semantic preferences and context-sensitivity. In *Words and intelligence II: Essays in honor of Yorick Wilks*, eds. K. Ahmad, C. Brewster, and M. Stevenson, 39–62. Springer: Dordrecht.
5. Darwin, C. 1872. *The expression of emotion in man and animals*. London: John Murray.
6. Eliot, T. S. 1922. *The waste land*. London: Faber.
7. Gratch, J., and S. Marsella. 2006. Evaluating a computational model of emotion. *Journal of Autonomous Agents and Multiagent Systems* (Special issue on the best of AAMAS 2004) 11(1):23–43.
8. Krippendorf, K. 1980. *Content analysis: An introduction to its methodology*. Beverly Hills, CA: Sage Publications.
9. Landauer, T.K., and Dumais, S.T. 1997. A solution to Plato's problem: The latent semantic analysis theory of acquisition, induction and representation of knowledge. *Psychological Review* 104:211–240.
10. Lee, M., and Y. Wilks. 1996. An ascription-based approach to speech acts. In Proceedings of International Conference on Computational Linguistics (COLING96), 699–704. Copenhagen.
11. Miller, G. A. 1985. WordNet: A dictionary browser. In Proceedings of the First International Conference on Information in Data, 25–28. Waterloo: University of Waterloo.
12. Ortony, A., G. L. Clore, and A. Collins. 1988. *The cognitive structure of emotions*. Cambridge: Cambridge University Press.
13. Ortony, A., and L. Fainsilber.1989. The role of metaphors in descriptions of emotions. In *Theoretical issues in natural language processing*, ed. Y. Wilks, 178–182. Hillsdale, NJ: Erlbaum.
14. Passonneau, R., T. Lippincott, T. Yano, and J. Klavans. 2008. Relation between agreement measures on human labeling and machine learning performance results from an art history domain. In Proceedings of Language Resources and Evaluation (LREC) Conference, Morocco.
15. Stone, P., D. Dunphy, S. Smith, D. Ogilvie, and Associates. 1966. *The General inquirer: A computer approach to content analysis*. Cambridge, MA: MIT Press.
16. Wiebe, J. 2000. Learning subjective adjectives from Corpora. In Proceedings of 17th National Conference on Artificial Intelligence (AAAI-2000), 735–740. Austin, TX.
17. Wilks, Y., J. Barnden, and J. Wang. 1991. Your metaphor or mine: Belief ascription and metaphor interpretation. In Proceedings of IJCAI91, 945–950.

Name Index

Subject Index